Nonlinear Opti
Investigations using DFT Calculations

S.Arulmani

Copyright © [2023]

Title : Nonlinear Optical Single Crystal Investigations using DFT Calculations

Author's : S. Arulmani

All rights reserved. No part of this publication may be reproduced, stored in a retrieval system, or transmitted in any form or by any means, electronic, mechanical, photocopying, recording, or otherwise, without the prior written permission of the publisher or author, except in the case of brief quotations embodied in critical reviews and certain other non-commercial uses permitted by copyright law.

This book was printed and published by [Publisher's: S.Arulmani] in [2023]

ISBN:

For permission to reproduce any of the material in this book.

TABLE OF CONTENTS

TABLE OF CONTENT
LIST OF FIGURES
LIST OF TABLES
PREFACE
LIST OF SYMBOLS AND ABBREVIATIONS
ABSTRACT IN TAMIL

CHAPTER No.	TITLE	PAGE No.
1	**INTRODUCTION TO EXPERIMENTAL AND COMPUTATIONAL STUDIES**	1
1.1	INTRODUCTION	1
1.2	SIGNIFICANCE OF SINGLE CRYSTAL GROWTH	1
1.3	CRYSTAL GROWTH	2
1.4	SLOW EVAPORATION TECHNIQUE	3
	1.4.1 AMINO ACIDS	3
1.5	SOLVENT SELECTION	4
	1.5.1 SOLUTION PREPARATION AND CRYSTAL GROWTH	5
	1.5.2 SEED PREPARATION	5
	1.5.3 AGITATION	5
1.6	NONLINEAR OPTICAL CRYSTALS	6
	1.6.1 THEORETICAL EXPLANATION OF NONLINEAR OPTICS	7
1.7	COMPUTATIONAL METHODS	10
	1.7.1 DENSITY FUNCTIONAL THEORY	10
	1.7.2 APPLICATIONS OF GAUSSIAN 09 W	11
1.8	THE SCOPE OF THE THESIS	11

CHAPTER NO.	TITLE	PAGE NO.
II	**EXPERIMENTAL AND COMPUTATIONAL STUDIES ON L-ARGININIUM HYDROGEN SQUARATE**	13
2.1	INTRODUCTION	13
2.2	SYNTHESIS AND GROWTH	14
2.3	RESULTS AND DISCUSSION	14
	2.3.1 SINGLE CRYSTAL X-RAY DIFFRACTION	14
	2.3.2 COMPUTATIONAL DETAILS	14
	2.3.3 MOLECULAR GEOMETRY	14
	2.3.4 VIBRATIONAL ASSIGNMENTS	17
	2.3.5 HYPERPOLARIZABILITY	22
	2.3.6 HOMO AND LUMO ANALYSIS	25
	2.3.7 THERMODYNAMICAL PROPERTIES	28
	2.3.8 OPTICAL STUDIES	28
	2.3.9 MULLIKEN POPULATION ANALYSIS	32
	2.3.10 NATURAL BONDING ORBITAL (NBO) ANALYSIS	39
	2.3.11 MICROHARDNESS TEST	39
	2.3.12 THERMAL ANALYSIS	41
	2.3.13 DIELECTRIC STUDIES	41
	2.3.14 PHOTOLUMINESCENCE ANALYSIS	46
	2.3.15 MOLECULAR ELECTROSTATIC POTENTIAL (MEP)	46
	2.3.16 SECOND HARMONIC GENERATION (SHG)	49
2.4	CONCLUSION	49
III	**EXPERIMENTAL AND COMPUTATIONAL STUDIES ON L- GLYCINIUM HYDROGEN SQUARATE**	50
3.1	INTRODUCTION	50
3.2	SYNTHESIS AND GROWTH	51
3.3	RESULTS AND DISCUSSION	51
	3.3.1 SINGLE CRYSTAL X-RAY DIFFRACTION	51

CHAPTER NO.		TITLE	PAGE NO.
	3.3.2	COMPUTATIONAL DETAILS	54
	3.3.3	MOLECULAR GEOMETRY	54
	3.3.4	VIBRATIONAL ASSIGNMENTS	58
	3.3.5	HYPERPOLARIZABILITY	62
	3.3.6	HOMO AND LUMO ANALYSIS	64
	3.3.7	THERMODYNAMICAL PROPERTIES	67
	3.3.8	OPTICAL STUDIES	67
	3.3.9	MULLIKEN POPULATION ANALYSIS	68
	3.3.10	NATURAL BONDING ORBITAL (NBO) ANALYSIS	77
	3.3.11	MICROHARDNESS TEST	77
	3.3.12	THERMAL ANALYSIS	80
	3.3.13	DIELECTRIC STUDIES	80
	3.3.14	PHOTOLUMINESCENCE ANALYSIS	84
	3.3.15	MOLECULAR ELECTROSTATIC POTENTIAL (MEP)	84
	3.3.16	SECOND HARMONIC GENERATION (SHG)	87
3.4	CONCLUSION		87
IV	**EXPERIMENTAL AND COMPUTATIONAL STUDIES ON L-ARGININIUM 5- NITROURACILATE**		**88**
4.1	INTRODUCTION		88
4.2	SYNTHESIS AND GROWTH		90
4.3	RESULTS AND DISCUSSION		89
	4.3.1	SINGLE CRYSTAL X-RAY DIFFRACTION	89
	4.3.2	COMPUTATIONAL DETAILS	89
	4.3.3	MOLECULAR GEOMETRY	92
	4.3.4	VIBRATIONAL ASSIGNMENTS	92
	4.3.5	HYPERPOLARIZABILITY	98
	4.3.6	HOMO AND LUMO ANALYSIS	99
	4.3.7	THERMODYNAMICAL PROPERTIES	103
	4.3.8	OPTICAL STUDIES	107

CHAPTER NO.		TITLE	PAGE NO.
	4.3.9	MULLIKEN POPULATION ANALYSIS	107
	3.3.10	NATURAL BONDING ORBITAL (NBO) ANALYSIS	114
	4.3.11	MICROHARDNESS TEST	114
	4.3.12	THERMAL ANALYSIS	116
	4.3.13	DIELECTRIC STUDIES	116
	4.3.14	PHOTOLUMINESCENCE ANALYSIS	121
	4.3.15	MOLECULAR ELECTROSTATIC POTENTIAL (MEP)	121
	4.3.16	SECOND HARMONIC GENERATION (SHG)	121
4.4		CONCLUSION	124
V		**EXPERIMENTAL AND COMPUTATIONAL STUDIES ON L-GLYCINIUM 5-NITROURACILATE**	125
5.1		INTRODUCTION	125
5.2		SYNTHESIS AND GROWTH	126
5.3		RESULTS AND DISCUSSION	126
	5.3.1	SINGLE CRYSTAL X-RAY DIFFRACTION	126
	5.3.2	COMPUTATIONAL DETAILS	126
	5.3.3	MOLECULAR GEOMETRY	129
	5.3.4	VIBRATIONAL ASSIGNMENTS	129
	5.3.5	HYPERPOLARIZABILITY	136
	5.3.6	HOMO AND LUMO ANALYSIS	136
	5.3.7	THERMODYNAMICAL PROPERTIES	140
	5.3.8	OPTICAL STUDIES	144
	5.3.9	MULLIKEN POPULATION ANALYSIS	144
	5.3.10	NATURAL BONDING ORBITAL (NBO) ANALYSIS	148
	5.3.11	MICROHARDNESS TEST	151
	5.3.12	THERMAL ANALYSIS	151
	5.3.13	DIELECTRIC STUDIES	153
	5.3.14	PHOTOLUMINESCENCE ANALYSIS	158

CHAPTER NO.	TITLE	PAGE NO.
	5.3.15 MOLECULAR ELECTROSTATIC POTENTIAL(MEP)	158
	5.3.16 SECOND HARMONIC GENERATION (SHG)	161
5.4	CONCLUSION	161
VI	**SUMMARY AND SUGGESTIONS FOR FUTURE WORK**	162
6.1	INTRODUCTION	162
6.2	SUMMARY	162
6.3	SUGGESTIONS FOR FUTURE WORK	166
	REFERENCES	167

LIST OF FIGURES

FIGURE No.	TITLE	PAGE No.
1.1	Second Harmonic Generation of Nonlinear Optical Crystal	8
2.1	Photograph of as grown LAHSQ single crystal	15
2.2	Atomic numbering system adapted for ab initio computations of LAHSQ molecule	18
2.3	Experimentally obtained FT-IR spectrum of LAHSQ	21
2.4	Theoretically simulated FT-IR spectrum of LAHSQ	21
2.5	HOMO – LUMO plot of LAHSQ at B3LYP/6-311G++ (d, p)	26
2.6	Variation of Enthalpy with temperature	29
2.7	Variation of Entropy with temperature	29
2.8	Variation of heat capacity with temperature	30
2.9	UV- Vis absorption spectrum of LAHSQ crystal	33
2.10	UV- Vis bandgap of LAHSQ crystal	33
2.11	UV-Vis transmission spectrum of LAHSQ crystal	34
2.12	UV- Vis reflective index of LAHSQ crystal	34
2.13	UV- Vis refractive index of LAHSQ crystal	35
2.14	UV-Vis extinction coefficient of LAHSQ crystal	35
2.15	Mulliken atomic charges of LAHSQ single crystal	36
2.16	Vickers hardness number Vs applied load of LAHSQ	42
2.17	log p Vs log d plot of LAHSQ	42
2.18	TG-DTA plot of LAHSQ crystal	43
2.19	Variation of dielectric constant of LAHSQ	44
2.20	Variation of dielectric loss of LAHSQ	44
2.21	Frequency dependence of AC Conductivity	45
2.22	Fluorescence emission spectrum of LAHSQ crystal	47
2.23	Molecular Electrostatic Potential (MEP) of LAHSQ crystal	48

3.1	Photograph of as grown LGHSQ single crystal	52
3.2	Atomic numbering system adapted for ab initio computations of LGHSQ molecule	55
3.3	Experimentally obtained FT-IR spectrum of LGHSQ	61
3.4	Theoretically simulated FT-IR spectrum of LGHSQ	61
3.5	HOMO – LUMO plot of LGHSQ at B3LYP/6-311G (d, p)	65
3.6	Variation of Enthalpy with temperature	69
3.7	Variation of Entropy with temperature	69
3.8	Variation of heat capacity with temperature	70
3.9	UV- Vis absorption spectrum of LGHSQ crystal	72
3.10	UV- Vis bandgap plot of LGHSQ crystal	72
3.11	UV-Vis transmission spectrum of LGHSQ crystal	73
3.12	UV- Vis reflective index of LGHSQ crystal	73
3.13	UV- Vis refractive index of LGHSQ crystal	74
3.14	UV-Vis extinction coefficient of LGHSQ crystal	74
3.15	Mulliken atomic charges of LGHSQ single crystal	75
3.16	Vickers hardness number Vs applied load of LGHSQ	79
3.17	log p Vs log d plot of LGHSQ	79
3.18	TG-DTA plot of LGHSQ crystal	81
3.19	Variation of dielectric constant with frequency of LGHSQ	82
3.20	Variation of dielectric loss with frequency of LGHSQ	82
3.21	Frequency dependence of AC Conductivity	83
3.22	Fluorescence emission spectrum of LGHSQ crystal	85
3.23	Molecular Electrostatic Potential (MEP) of LGHSQ crystal	86
4.1	Photograph of as grown LA5N single crystal	90
4.2	Atomic numbering system adapted for ab initio computations of LA5N molecule	93
4.3	Experimentally obtained FT-IR spectrum of LA5N	96
4.4	Theoretically simulated FT-IR spectrum of LA5N	96
4.5	HOMO – LUMO plot of LA5N at B3LYP/6-311++G (d, p)	101

4.6	Variation of Enthalpy with temperature	104
4.7	Variation of Entropy with temperature	104
4.8	Variation of heat capacity with temperature	105
4.9	UV- Vis absorption spectrum of LA5N crystal	108
4.10	UV- Vis bandgap plot of LA5N crystal	108
4.11	UV-Vis transmission spectrum of LA5N crystal	109
4.12	UV- Vis reflective index of LA5N crystal	109
4.13	UV- Vis refractive index of LA5N crystal	110
4.14	UV-Vis extinction co-efficient of LA5N crystal	110
4.15	Mulliken atomic charges of LA5N single crystal	111
4.16	Vickers hardness number Vs applied load of LA5N	117
4.17	log P Vs log d plot of LA5N	117
4.18	TG-DTA plot of LA5N crystal	118
4.19	Variation of dielectric constant of LA5N	119
4.20	Variation of dielectric loss of LA5N	119
4.21	Frequency dependence of AC Conductivity	120
4.22	Fluorescence emission spectrum of LA5N crystal	122
4.23	Molecular Electrostatic Potential (MEP) of LA5N crystal	123
5.1	Photograph of as grown LGY5N single crystal	127
5.2	Atomic numbering system adapted for ab initio computations of LGY5N molecule	130
5.3	Experimentally obtained FT-IR spectrum of LGY5N	133
5.4	Theoretically simulated FT-IR spectrum of LGY5N	133
5.5	HOMO – LUMO plot of LGY5N at B3LYP/6-311++G (d, p)	138
5.6	Variation of Enthalpy with temperature	141
5.7	Variation of Entropy with temperature	141
5.8	Variation of heat capacity with temperature	142
5.9	UV- Vis absorption spectrum of LGY5N crystal	145
5.10	UV- Vis bandgap plot of LGY5N crystal	145
5.11	UV-Vis transmission spectrum of LGY5N crystal	146

5.12	UV- Vis reflective index of LGY5N crystal		146
5.13	UV- Vis refractive index of LASQ crystal		147
5.14	UV-Vis extinction coefficient LGY5N of crystal		147
5.15	Mulliken atomic charges of LGY5N single crystal		149
5.16	Vickers hardness number Vs applied load of LGY5N		154
5.17	log p Vs log d plot of LGY5N		154
5.18	TG-DTA plot of LGY5N crystal		155
5.19	Variation of dielectric constant of LGY5N		156
5.20	Variation of dielectric loss of LGY5N		156
5.21	Frequency dependence of AC Conductivity		157
5.22	Fluorescence emission spectrum of LGY5N crystal		159
5.23	Molecular Electrostatic Potential (MEP) of LGY5N crystal		160

LIST OF TABLES

TABLE No.	TITLE	PAGE No.
2.1	Crystal Parameters of LAHSQ single crystal	16
2.2	Selected bond lengths of LAHSQ molecule	19
2.3	Selected bond angles of LAHSQ molecule	20
2.4	The electric dipole moment μ, the average polarizability α_{tot} and first hyperpolarizability β_{tot} for LAHSQ molecule	24
2.5	Calculated electronic and energies of LAHSQ using B3LYP/6-311++G (d, p) level	27
2.6	Thermodynamic properties at different temperatures by B3LYP level for LAHSQ	31
2.7	Mulliken atomic charges of LAHSQ single crystal	37
2.8	Second order perturbation theory analysis of Fock Matrix in NBO for LAHSQ with 6-311++ G (d, p) basis set	40
3.1	Crystal parameters of LGHSQ single crystal	53
3.2	Selected bond lengths of LAHSQ molecule	56
3.3	Selected bond angles of LGHSQ molecule	57
3.4	The electric dipole moment μ, the average polarizability α_{tot} and first hyperpolarizability β_{tot} for LGHSQ molecule	63
3.5	Calculated electronic and energies of LGHSQ using B3LYP/6-311++G (d, p) level	66
3.6	Thermodynamic properties at different temperatures by B3LYP level for LGHSQ	71
3.7	Mulliken atomic charges of LGHSQ single crystal	76
3.8	Second order perturbation theory analysis of Fock Matrix in NBO for LGHSQ with 311++ G (d, p) basis set	78
4.1	Crystal parameters of LA5N single crystal	91
4.2	Selected bond lengths of LA5N molecule	94
4.3	Selected bond angle of LA5N molecule	95
4.4	The electric dipole moment μ, the average polarizability α_{tot} and first hyperpolarizability β_{tot} for LA5N molecule	100

4.5	Calculated electronic and energies of LA5N using B3LYP/6-311++G (d, p) level	102
4.6	Thermodynamic properties at different temperatures by B3LYP level for LA5N	106
4.7	Mulliken atomic charges of LA5N single crystal	112
4.8	Second order perturbation theory analysis of Fock Matrix in NBO for LA5N with 6-311++ G (d, p) basis set	115
5.1	Crystal parameters of LGY5N single crystal	128
5.2	Selected bond lengths of LGY5N molecule	131
5.3	Selected bond angles of LGY5N molecule	132
5.4	The electric dipole moment μ, the average polarizability α_{tot} and first hyperpolarizability β_{tot} for LGY5N molecule	137
5.5	Calculated electronic and energies of LGY5N using B3LYP/6-311++G (d, p) level	139
5.6	Thermodynamic properties at different temperatures by B3LYP level for LGY5N	143
5.7	Mulliken atomic charges of LGY5N single crystal	150
5.8	Second order perturbation theory analysis of Fock Matrix in NBO for LGY5N with 6-311++ G (d, p) basis set	152

PREFACE

Crystal growth plays an important role in basic research and technologies, such as laser technology, optical communication and optical data storage. Semiconductor based devices, transducers, infrared detectors, solid state lasers, frequency converters, optical devices based on nonlinear optics, piezo electrics and acousto-optics essentially need perfect single crystals. Hence, growth of good quality single crystals becomes inevitable for further research and technology. The growth of single crystal is primarily based on the ease of use, nature of the starting materials and their physicochemical properties. Nonlinear optics (NLO) is a new branch of science that began soon after the development of laser. After the discovery of NLO effect, scientists immediately renowned that any practical applications of NLO would depend on the development of new materials. These optical materials are playing important role in the novel applications, such as frequency conversion by laser harmonic crystals, light amplitude and phase modulation by electro-optic crystals and phase conjugation by photorefractive crystals. Hence the search for new NLO materials is still active and interesting for the technological development. Crystals possessing NLO properties change the propagation characteristics (phase, frequency, amplitude, polarization) of the incident light with applications in communication, high power laser generation by frequency conversion, optical switching and advanced optical devices. New device concepts and materials developments are the two key factors for future success. Photonics is one among several contenders to replace or at least displace electronics for computing and information processing. The advantages of photonic devices, using NLO processes, over electronic devices are faster in speed, owing to the use of photons instead of electrons and higher bandwidth capacity. In recent years amino acid based NLO crystals with enhanced linear and nonlinear optical properties were recognized. Hence an attempt was made to grow and characterize amino acid based NLO crystals. The procedure and results are summarized in this thesis.

The present thesis is aimed towards the growth and characterization of organic nonlinear optical (NLO) of L- Argininium Hydrogen Squarate (LAHSQ), L- Glycinium Hydrogen Squarate (LGHSQ), L-Argininium 5- Nitrouracilate (LA5N), L- Glycinium 5- Nitrouracilate (LGY5N) single crystals.

The thesis entitled *"Experimental Investigations and Density Functional Theory Calculations On LAHSQ, LGHSQ, LA5N and LGY5N Nonlinear Optical Single Crystals Crystal"* consists of six chapters.

Chapter-I deals with the basic details regarding the crystal and crystal growth of nonlinear optical (NLO) with a basic explanation. An overview about nonlinear optics, nonlinear optical crystals and their applications are discussed. Followed by this, an outline to Density Functional Theory (DFT) and its applications in exploring the theoretical aspects are also explained in detail.

Chapter II to V gives a detailed account of the studies conducted on the grown single crystals. These chapters also give an account of the various characterization techniques employed to evaluate the prospects of title materials for NLO applications. XRD analysis, Density Functional Theory (DFT) computations such as bond lengths and bond angles were obtained for all the materials. Hyperpolarizability and HOMO - LUMO analyses, Thermodynamical characters, MEP, Mulliken charges and NBO's were performed with the B3LYP/6-311++G (d, p) basis sets. Theoretical and Experimental FTIR analysis were done to identify various functional groups present in the materials. Optical parameters such as band gap, reflectance (R), refractive index (n) and extinction coefficient (K) were obtained for the materials. Thermal analysis of the materials revealing the stability nature of the title crystal. Mechanical strength and SHG efficiency were done for all the materials to understand its suitability in NLO applications. Dielectric studies were carried out to recognize the electrical response of the crystals.

Chapter VI gives a summary of the investigations carried out on these four crystals, along with the suggestion for future study. The results of the above investigations have been published / accepted in International Journals and were also presented in various National/International seminars/conferences.

LIST OF SYMBOLS AND ABBREVATIONS

α	Polarizability
β	First order hyperpolarizability
ε_0	Permittivity of free space
ε_r	Dielectric constant
η	Conversion efficiency
λ	Wavelength
ω	Electrophilicity index
$\chi^{(1)}$	Linear susceptibility
$\chi^{(2),(3)}$	Nonlinear susceptibilities
μ	Dipole moment
Å	Angstrom
a.u	Astronomical unit
°C	Degree Celsius
μ_0	Permeability of free space
S	Softness
EA	Electron Affinity
Eg	Energy Band gap (eV)
eV	Electron volts
mm	Millimetre
ΛK	Phase mismatching
μm	Micrometer
μv	Microvolt
B3LYP	Becke–Lee–Yang–Parr hybrid exchange-correlation three-parameter functional

DFT	Density Functional Theory
FT-IR	Fourier Transform Infrared
Hv, VHN	Vickers hardness number
IR	Infrared
FMO	Frontier Molecular Orbital
LAHSQ	L-Argininium Hydrogen Squarate
LGHSQ	L-Glycinium Hydrogen Squarate
LA5N	L-Argininium 5-Nitrouracilate
LGY5N	L-Glycinium 5-Nitrouracilate
n	Work hardening coefficient
NLO	Non-Linear Optics
SHG	Second Harmonic Generation
UV	Ultraviolet
TGA	Thermogravimetric Analysis
SCXRD	Single Crystal X-Ray Diffraction
KDP	Potassium Dihydrogen Phosphate
HOMO	Highest Occupied Molecular Orbit
LUMO	Lowest Unoccupied Molecular Orbital
ICT	Intermolecular Charge Transfer
IP	Ionization Potential
PL	Photoluminescence

CHAPTER I

INTRODUCTION TO EXPERIMENTAL AND COMPUTATIONAL STUDIES

1.1 INTRODUCTION

Modern technology is seeing a rapid change and its reflection on mankind is splendid. The technological development to a larger extent is dependent on the development of crystal growth. Crystal growth has prominent role to play in present era of rapid technical and scientific advancement where the application of crystals has unbounded limits. A single crystal is periodic array of atoms arranged in a three-dimensional structure with equally repeated distances in a given direction. Crystals have wide applications especially in the fields of communication, laser technology, thermal imaging and electronics. Modern technology is based on the single crystals of nonlinear optical ferroelectric, semiconductor, superconductor and acousto-optic materials. Significant advancement in crystal growth technology has allowed the development of many excellent crystals to meet the ever-growing applications in laser, optical communications and data storage technology. Hence, growth of single crystal has become inevitable for the further research in material science and crystal technology.

Frequency conversion is an important and popular technique for extending the useful range of laser. This has led to production of various devices such as harmonic generator optical parametric oscillators, electronic modulators and amplifiers for high power lasers. Recent efforts are focused on the development of new and efficient frequency materials for such applications. The search of new material is primarily focused on increasing nonlinearity. With progress in crystal growth and technology, materials having attractive nonlinear properties are being discovered. This has enabled the commercial development of single crystals with promising NLO properties. To enable a material to be potentially useful for NLO application, the material should be available in the bulk single crystal form. Besides this, aqueous solution is serving as the most important to general progress in understanding many fundamental concepts of crystallization.

1.2 SIGNIFICANCE OF SINGLE CRYSTAL GROWH

The significance of the beauty and rarity of crystal now well knitted with their symmetry, molecular structure and their purity and the physicochemical environment of their formation. These characteristics endow crystal with unique physical and chemical properties, which have transformed electronic industries for the benefit of human society. The strong influence of single crystal in precent-day technology is evident from recent advantages in the field of semi-conductors, transducers, Infrared detectors, ultrasonic amplifiers, solid state lasers, non-linear optics, piezoelectric, acousto-optics, photosensitive material and crystalline thin films for microelectronics and the computer applications (Dhanaraj G et. al. 2010). Prior to commercial growth or production of crystals, man depended only on the availability of the natural crystals for both jewellery and devices. But today the of artificially grown crystal is growing exponentially for a variety of applications.

Crystal growth is basically a process of arranging atoms, ions, molecules or molecular assemblies into regular three-dimensional periodic arrays. However real crystal is never perfect, mainly due to the presence of different kind of local disorder and long-range imperfection such as dislocations. Hence the ultimate aim of the crystal grower is to produce perfect single crystal of desired size and shape and to characterize them in order to understand the purity, quality and perfection. Accordingly, crystal techniques and characterization tools have advanced greatly in recent years. This has facilitated the growth of a large variety of a technologically important single crystal. Crystal growth has been treated as important branch of material science leading to the formation of technologically important material of different sizes (Buckley H. E et. al. 1951and Mullin J.W et. al. 1976). Thus, the growth of single crystal and their characterization towards device fabrication have assumed great impetus due to their significance in both academic research and applied research.

1.3 CRYSTAL GROWTH

Crystal growth is a heterogeneous or homogeneous chemical process involving solid, liquid or gas, whether individually or together, to form a homogeneous solid substance having three-dimensional atomic arrangement. Based on the phase transformation process, crystal growth techniques are classified as given below:

Growth from solid ⟹ solid - solid phase transformation
Growth from liquid ⟹ liquid - solid phase transformation
Growth from vapour ⟹ vapour - solid phase transformation

Crystal growth is the process of arranging atoms or molecules in a fluid or solution state into an ordered solid state. Crystals grow by the ordered deposition of material from the fluid or solution state to the surface of the crystal.

The methods of growing crystals are very wide and mainly dictated by the characteristics of the material and its size (Buckley H. E et. al. 1951 and Mullin J. W et. al. 1976). An efficient process is the one, which produces crystals adequate for their use at minimum cost. The growth method is essential because it suggests the possible impurity and other defect concentrations. Choosing the best method to grow a given material depends on material characteristics. In the present work, single crystals of L-Argininium hydrogen squarate (LAHSQ), L-Glycinium hydrogen Squarate (LGHSQ), L-Argininium 5-Nitrouracilte (LA5N) and L-Glycinium 5-Nitrouracilate (LGY5N) are grown employing slow evaporation technique, hence the fundamentals of this technique are outlined in this chapter.

1.4 SLOW EVAPORATION TECHNIQUE

In this method, the saturated solution is kept at a particular temperature and provision is made for evaporation. If the solvent is non-toxic like water, it is easy to allow evaporation into the atmosphere. The evaporation technique has the advantage to grow crystal at a fixed temperature. This technique needs a vessel for keeping the solution in which the crystals grow. The height, radius and volume of the vessel are properly chosen to achieve the required crystal size. This method is perhaps the easiest method of growing crystals and is cheap enough to permit many parallel attempts to be made. In a typical procedure, a saturated solution heated slightly above its saturation temperature is poured into a screw-cap jar and a seed tied to a piece of thread is introduced to achieve the growth of the crystals.

1.4.1 AMINO ACIDS

An amino acid mainly contains a carboxylic group and an amino group. Amino acids form the building blocks of the protein structure. Amino acids are classified on the basis of their acidic/basic nature, chemical structure, nutritional requirement and relative position of amino acid and carboxylic acid groups. Based on acidic or basic nature, the amino acids are classified into neutral amino acids, acidic amino acids and basic amino acids. Amino acids containing one amino and one carboxylic acid group are known as neutral amino acids. Some examples of neutral amino acids are glycine, alanine and serine etc. Amino acids containing one amino and two carboxylic acid groups are known as acidic amino acids. The second carboxylic acid group may be either a free acidic group or with an amino group. Examples of this type are aspartic acid, asparagine, glutamic acid etc. Amino acids containing one carboxylic acid group and two amino or imino groups are known as basic amino acids. Example of this type is lysine. Based on the chemical structure amino acids are classified into (i) amino acids with aliphatic side chains; Examples: glycine, alanine, valine, (ii) amino acids containing hydroxyl groups; Examples: Serine, Threonine, (iii) sulphur containing amino acids; Examples: cysteine, methionine, (iv) aromatic amino acid; Examples: phenylalanine, tyrosine, and (v) imino acids; Examples: proline, hydroxy proline etc. Based on nutritional requirement, the amino acids are classified into two types viz. essential amino acids and non-essential amino acids. The amino acids which cannot be synthesized by the body and therefore need to be supplied through the diet are called essential amino acids. Examples: valine, lysine, arginine. The amino acids which can be synthesized by the body and therefore need not be supplied through the diet are called non-essential amino acids. Examples: glycine, alanine, serine etc. Amino acids are also classified based on the relative position of amino and carboxylic acid groups. Amino acids are classified as α, β and γ-amino acids on the basis of the relative position of amino group and carboxylic acid group. Examples for α-amino acids are glycine, alanine. Examples for β-amino acids are β-alanine and examples for γ-amino acids are γ-amino butyric acid and γ-glycine. In the zwitterionic state the amino acids are electrically neutral and bear no net charge. Therefore, in the zwitterionic state they do not migrate under the applied electric field and possess least solubility. In strongly acidic medium the amino acids exist as cations and migrate towards cathode. In strongly alkaline medium they exist as anions and migrate towards anode. Among organic class of NLO materials, amino acids exhibit some specific features such as molecular chirality weak Vander Waals hydrogen bonds, absence of strongly conjugated bonds wide transparency range in the visible and ultraviolet regions, amenability for synthesis multi-functional substitution, higher resistance to optical damage and manoeuvrability for device application etc., (Arulmani S. et. al. 2018).

1.5 SOLVENT SELECTION

A solution is homogeneous mixture of a solute in a solvent. Solute is the component, which is present in a smaller quantity and the one which gets dissolved in the solvent. For a given solute, there may be different solvents. The solvent must be chosen taking into account the following factors to grow crystals from solution. A solvent of choice is the one with,

- a good solubility for the given solute
- a good temperature coefficient of solute solubility
- less viscosity
- less volatility
- less corrosion and non-toxicity and
- cost effective

1.5.1 SOLUTION PREPARATION AND CRYSTAL GROWTH

The solubility curve can be traced in different temperatures by repeating the above procedure. The greater solubility of substances increases with temperature. The clear solution, saturated at the desired temperature, is taken into a growing vessel. A tiny crystal suspended into the solution is used for saturation testing. By changing the temperature, either the occurrence of growth or dissolution is not determined. For the growth of bulk single crystal, the test seed is replaced by a good quality seed crystal. All unwanted nuclei and surface damaged seeds are removed. Growth is initiated after saturation. Solvent evaporation can also help to stimulate growth. The quality of the grown crystal depends on the (a) nature of seed, (b) cooling rate employed and (c) agitation of the solution.

1.5.2 SEED PREPARATION

Seed crystals are prepared by self-nucleation under slow evaporation from a saturated solution. Seeds of good visual quality, free from any inclusion and imperfections are chosen for growth. Since, strain free refracting of the seed crystal results in low dislocation content, a few layers of the seed crystal are dissolved before initiating the growth. Defects present in an imperfect seed propagate into the bulk of the crystal, which decreases the quality of the crystal. Hence, seed crystals are prepared with care. The quality of the bulk crystal is usually slightly better than that of the seed.

1.5.3 AGITATION

To have a regular and even growth, the level of super saturation has to be maintained equally around the surface of the growing crystal. An uneven growth leads to localized stresses at the surface generating imperfection in the bulk crystals. Moreover, the concentration gradients that exist in the growth vessels at different faces of the crystal cause fluctuations in super saturation, seriously affecting the growth rate of individual faces. The gradient at the bottom of the growth vessel exceeds the metastable zone width, resulting in spurious nucleation. The degree of formation of concentration gradients around the crystal depends on the efficiency of agitation of the solution. This is achieved by agitating the saturated solution in either direction at an optimized speed using a stirrer motor.

1.6 NONLINEAR OPTICAL CRYSTALS

When an electromagnetic wave propagates through a material, the electric field induces a time varying polarization in the medium. In a linear medium the magnitude of the induced polarization is proportional to the amplitude of the electric field. In nonlinear optical materials this is not the case, especially when the amplitude of the electric field is large. An intense light beam, such as a laser beam, propagating through a nonlinear optical material will produce effects that cannot be observed with weak light beams. The basic requirement for a nonlinear crystal is that it should have an excellent optical quality. Hence, it is necessary to grow single crystal specimen of good optical quality. Thus, in many cases the search for new and better non-linear optical materials is very largely a crystal growing effort. It is realized that the requirements on optical quality for nonlinear optics are more stringent than even the most exciting requirements on optical quality for materials used in linear optics. For a successful device fabrication, it is essential for a material to meet a number of criteria in crystal properties. The relevant issues include reliable crystal growth techniques, ready availability, optical non-linearity, birefringence, moderate to high transparency and optical homogeneity for high conversion efficiency, mechanical strength, chemical stability, cutting and polishing feasibility, phase matching bandwidth, fracture toughness, thermo-mechanical properties for high average power, high laser damage threshold and brittleness index for lifetime and system capability which are essential to make an ideal device. The phenomenon of the second harmonic generation (SHG) in a quartz crystal, was first reported by (Franken et. al. 1961), which led to the development of many NLO materials, such as lithium niobate ($LiNbO_3$), potassium niobate ($KNbO_3$), barium titanate ($BaTiO_3$), potassium titanyl phosphate ($KTiOPO_4$), potassium dihydrogen phosphate (KH_2PO_4), lithium iodate, β-barium borate (BBO) and numerous organic and semi-organic materials. An ideal nonlinear optical material should possess the following characteristics:

- Large nonlinear figure of merit for frequency conversion,
- High laser damage threshold,
- Fast optical response time,
- Wide phase matching angle,
- Architectural flexibility for molecular design and morphology,
- Ability to process into crystals, thin films,
- Optical transparency,
- Easy fabrication,
- Non-toxicity and good environmental stability and
- High mechanical strength and thermal stability.

1.6.1 THEORETICAL EXPLANATION OF NONLINEAR OPTICS

The explanation of nonlinear effects lies in the way in which a beam of light propagates through a solid. The nuclei and associated electrons of the atoms in the solid form electric dipoles. The electromagnetic radiation interacts with these dipoles causing them to oscillate, which by the classical laws of electromagnetism, results in the dipoles themselves acting as sources of electromagnetic radiation. If the amplitude of vibration is small, the dipoles emit radiation of the same frequency as the incident radiation. As the intensity of the incident radiation increases, the relationship between irradiance and amplitude of vibration becomes nonlinear resulting in the generation of harmonics in the frequency of radiation emitted by the oscillating dipoles. Thus, frequency doubling or second harmonic generation (SHG) and indeed higher order frequency effects occur as the incident intensity is increased. In a nonlinear medium the induced polarization is a nonlinear function of the applied field. A medium exhibiting SHG is a crystal composed of molecules with asymmetric charge distributions arranged in the crystal in such a way that a polar orientation is maintained throughout the crystal. At very low fields, the induced polarization is directly proportional to the electric field.

$$P = \varepsilon_0 \chi E \qquad (1.1)$$

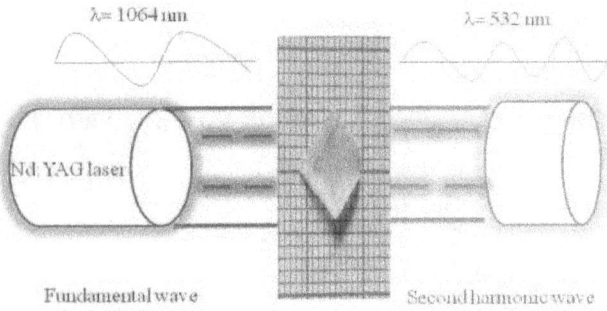

Fig. 1.1 Second Harmonic Generation of Nonlinear Optical Crystal

Where χ is the linear susceptibility of the material, E is the electric field vector, ε_0 is the permittivity of free space. At high fields, polarization becomes independent of the field and the susceptibility becomes field dependent. Therefore, this nonlinear response is expressed by writing the induced polarization as a power series in the field.

$$P = \varepsilon_0 \chi^{(1)} E + \chi^{(2)} E \cdot E + \chi^{(3)} E \cdot E \cdot E + \ldots \tag{1.2}$$

In nonlinear terms, product of two or more oscillating fields gives oscillation at combination of frequencies and therefore the above equation can be expressed in terms of frequency as:

$$P(-\omega_0) = \varepsilon_0 \chi^{(1)}(-\omega_0; \omega_1) \cdot E(\omega_0) + \chi^{(2)}(-\omega_0; \omega_1, \omega_2) \cdot E\omega_1 \cdot E\omega_2 + \chi^{(3)}(-\omega_0; \omega_1, \omega_2, \omega_3) \cdot E\omega_1 \cdot E\omega_2 \cdot E\omega_3 + \ldots \tag{1.3}$$

Where $\chi^{(2)}$, $\chi^{(3)}$ .... are the nonlinear susceptibilities of the medium. $\chi^{(1)}$ (is the linear term responsible for material's linear optical properties like refractive index, dispersion, birefringence and absorption $\chi^{(2)}$ is the quadratic term which describes second harmonic generation in non-centrosymmetric materials. $\chi^{(3)}$ is the cubic term responsible for third harmonic generation, stimulated Raman scattering, phase conjugation and optical instability. Hence the induced polarization is capable of multiplying the fundamental frequency to second, third and even higher harmonics. If the molecule or crystal is centrosymmetric then $\chi^{(2)} = 0$. If a field +E is applied to the molecule (or medium), equation 1.3 predicts that the polarization induced by the first nonlinear term is predicted to be +E2, yet if the medium is centrosymmetric the polarization should be –E2. This contradiction can only be resolved if $\chi^{(2)} = 0$ in centrosymmetric media. If the same argument is used for the next higher order term, +E produces polarization +E3 and – E produces – E3, so that $\chi^{(3)}$ is the first non-zero nonlinear term in centrosymmetric media. In second harmonic generation, the two input wavelengths are the same $2\omega_1 = \omega_2$ or ($\lambda_1 = 2\lambda_2$). During this process, a polarized wave at the second harmonic frequency $2\omega_1$ is produced. The refractive index, n_1 is defined by the phase velocity and wavelength of the medium. The energy of the polarized wave is transferred to the electromagnetic wave at a frequency ω_2. The phase velocity and wavelength of this electromagnetic wave are determined by n_2, the refractive index of the doubled frequency. To obtain high conversion efficiency, the phase vectors of input beams and generated beams are to be matched.

$$\Delta k = - \frac{2\pi}{(n2-n1)} = 0 \qquad (1.4)$$

Where Δk represents the phase–mismatch. The phase–matching can be obtained by angle tilting, temperature tuning or other methods. Hence, to select a nonlinear optical crystal, for a frequency conversion process, the necessary criterion is to obtain high conversion efficiency. The conversion efficiency η is given by

$$\eta = PL^2 \, d_{eff} \frac{\sin \Delta KL^2}{\Delta KL} \qquad (1.5)$$

Where d_{eff} is the effective nonlinear coefficient, L is the crystal length, P is the input power density and Δk is the phase – mismatching. In general, higher power density, longer crystal, large nonlinear coefficients, and smaller phase mismatching will result in higher conversion efficiency. Also, the input power density must be lower than the damage threshold of the crystal.

1.7 COMPUTATIONAL METHODS

Computational chemistry also named molecular modelling is a set of explore systems on computer. Computational chemistry can be pronounced as chemistry performed using computer rather than chemicals. Computers are used to produce molecular properties rather than doing experiments (Boopathi. K et. al. 2018). The molecules simulated by computer will be helpful in synthesis the molecule in laboratory. There are also understandings into molecular bonding, obtained from the results of simulation that never be found from any investigational method. Many experimental chemists depend on electronic structural configuration of the element to foretell stable structure of the molecule and physiochemical properties. Then above said properties are used to research the ingredients that are highly challenging to find or very affluent to purchase, it helps to make expectation before running the actual experiment so can be better arranged for making observations (Indumathi. P et. al. 2018). The results presented in this dissertation were performed using Density Functional Theory and the same is described below.

1.7.1 DENSITY FUNCTIONAL THEORY

Density functional theory (DFT) is a methodology in finding a solution to the basic equations of quantum mechanically treated atoms and molecules. For three decades Density Functional Theory has been dominant method for the simulation of periodic systems. This theory has been developed more recently more than basic computational methods. It's different from other than ab initio method because de Broglie's wave concept is not used to pronounce the solid. The population of electrons is used instead of wave function (Rajagopalan. N. R et. al. 2018). There exist three types of calculation, first one-local density approximation (LDA) the second is-a correction in potential variation - provides additional accurate geometrical coordinates and the third one -hybrids (a blend of DFT and HF methods) - contribute extra precise geometries. The whole field of Density Functional Theory breaks on two fundamental mathematical theorems proved by Kohn and Hohenberg, and the derivation of a set of equations by Kohn and Sham. Theorem one states that here subsists a one-to-one mapping between the lowest state wave function and the minimum energy state electron density. Theorem two states that the lower most state electron density that curtails the energy functional. Local density approximation (LDA) is the unpretentious guess to the whole problem it is only cantered on the electron concentration for closed shell system. Local spin density approximation is only for high spin system. Thus, adapting electron concentration is more advantageous. The B3LYP method with basis sets of 6-311++G (d, p) or larger is the choice for many organic molecule calculations. The same is used here (Devi P. et. al. 2018 and Faizan M et. al. 2018).

1.1.1 APPLICATIONS OF GAUSSIAN 09 W

Gaussian 09W is one in a series of programs designed, improved and expanded over many years. Gaussian 09W was used exclusively for the calculations in this work and was utilized to visualize optimized geometries and calculated frequencies from Gaussian output files which are the basic parameters required for the interpretation of geometrical molecular structures and computing the nonlinear optical parameters, such as polarizability, dipole moment and hyperpolarizability and other parameters.

1.2 THE SCOPE OF THE THESIS

The ever-increasing demand for highly efficient nonlinear optical (NLO) crystals for visible and ultraviolet regions is extremely important for laser and material processing. In this context, the molecular design and growth of single crystal materials suitable for such requirements, assumes centre stage. Amino acid based NLO active L-Argininium hydrogen squarate (LAHSQ), L-Glycinium hydrogen Squarate (LGHSQ), L-Argininium 5-Nitrouracilte (LA5N) and L-Glycinium 5-Nitrouracilate (LGY5N) have attracted attention as promising materials keeping this in view, attempts are made to grow and study LAHSQ, LGHSQ, LA5N and LGY5N single crystals.

The present investigation is aimed at:

- Synthesizing high purity materials, like L-Argininium hydrogen squarate (LAHSQ), L-Glycinium hydrogen Squarate (LGHSQ), L-Argininium 5-Nitrouracilte (LA5N) and L-Glycinium 5-Nitrouracilate (LGY5N) using room temperature solution growth method.
- Confirmation of the structure and quality of the grown single crystals done by X-ray diffraction technique.
- Density Functional Theory investigations employed to understand molecular behaviour of the materials.
- Confirming the protonation of the amino group and the mode of vibration of different molecular group using FT-IR analysis.
- Determining the thermal stability of the grown crystals to assess their suitability in device fabrication.
- Understanding the mechanical behaviour of the grown crystals and analysing the dielectric property of the grown crystals with respect to frequency and temperature.
- Analysing the grown crystal for its optical absorption, NLO property and also the photoluminescence nature on the grown single crystals.

CHAPTER II

EXPERIMENTAL AND COMPUTATIONAL STUDIES ON L-ARGININIUM HYDROGEN SQUARATE (LAHSQ) SINGLE CRYSTAL

2.1 INTRODUCTION

Amino acid crystals usually display large nonlinear optical (NLO) response and are potential candidates for applications in the emerging areas of photonics (Badan et. al. 1993). Molecules that show asymmetric polarization induced by electron donor and acceptor groups are responsible for electro optic and NLO properties (Prasad et. al. 1991). Over the past two decades much attention has been paid to the search of novel high quality NLO materials that can generate high second order optical nonlinearities (Aggarwal M.D et. al. 2003; Razzetti M. et. Al. 2002; Ardle M.C et. al. 1974; Danushkodi M. and and Ramajothi 2004). The search for new advanced materials is an important area of contemporary research in numerous disciplines of science and development of many new technologies. The Nonlinear optical (NLO) crystals have become of great research interest and importance in the recent years for the fabrication of devices used in the field of telecommunication, optical signal processing, optical switching and Photonics. Now a day, various growth methods and apparatus have been continuously developed to improve the quality and growth rate. Compared to the other techniques, the slow evaporation technique is mostly used in several types of crystals. Organic crystals in terms of NLO properties possess advantages when compared with inorganic counterparts. Organic materials allow their fine tuning of their chemical structure and properties for the desired NLO properties.

In the present investigation, a desirable organic nonlinear optical single crystal of L-Argininium Hydrogen Squarate (LAHSQ) was synthesized using slow evaporation technique from aqueous solution. The density functional theory (DFT) is an actual tool to obtained the exact results of dipole moment, polarizabilities, hyperpolarizability and electronic structure of the LAHSQ crystal. To search the theoretical and experimental uniformity, quantum chemical calculations were carried out with complete geometry optimizations of the molecule. The grown crystals have been subject to different characteristics such as single crystal XRD, FTIR, UV-Vis, PL, TG/DTA, dielectric study, Microhardness study, Photoluminescence studies and NLO properties of also LAHSQ has been reported and the obtained results are elaborately discussed.

2.2 SYNTHESIS AND GROWTH

L-Argininium Hydrogen Squarate (LAHSQ) Single crystals were grown by the slow evaporation technique at room temperature method. High pure L-Arginine – $C_6H_{14}N_4O_2$ (Sigma–Aldrich, 99%) and Squaric acid $C_4H_2O_4$ (Sigma–Aldrich, 99%) were taken in equimolar ratio (1:1) and mixed using a magnetic stirrer for 5-6 hours to ensure homogeneous concentration in the entire volume of the solution. The recrystallization processes help to reduce the impurity present in the material. After that, the solution was filtered twice and placed in a dust free atmosphere. The optically good quality gray colour single crystal (7x4x3mm^3) was obtained in the period of 30-40 days. The grown LAHSQ single crystal is shown in Fig. 2.1.

2.3 RESULTS AND DISCUSSION

2.3.1 SINGLE CRYSTAL X-RAY DIFFRACTION

The structural property of LAHSQ was studied by Single Crystal X-ray diffraction technique. Single crystal X-ray diffraction data were recorded using ENRAF NONIUS CAD4 X-ray Diffractometer with M_oK_α radiation ($\lambda = 0.71073$Å) to obtain the lattice parameters and space group. The single crystal X-ray diffraction data shows that the grown LAHSQ crystal belongs to the P1 space group with triclinic crystal system. The space group shows that the crystal is non-centrosymmetric which is a basic condition for SHG applications. The unit cell parameters of LAHSQ are a =5.1301Å, b =8.3224Å, c =14.9005Å, V=621 (Å^3) and the crystallographic data and refinement details are given in Table 2.1.

2.3.2 COMPUTATIONAL DETAILS

The molecular structure optimization of the title compound and corresponding vibrational frequencies were calculated using the Density Functional Theory (DFT) with Beckee-3-Lee-Yag-Parr (B3LYP) combined with 6-311++G (d, p) basis set. The fundamental vibrational frequencies, IR intensity were calculated using the Gaussian 09 package (Frisch et al 2004). By combining the results of the GAUSSVIEW program with symmetry considerations, vibrational frequency assignments were made with a high degree of accuracy.

2.3.3 MOLECULAR GEOMETRY

The atomic numbering of elements in the molecular structure is presented in Fig. 2.2. The structural parameters like bond length and bond angle are obtained by using the B3LYP were

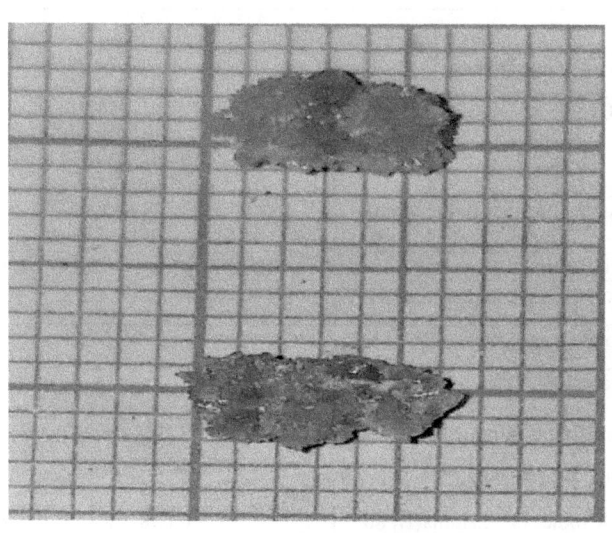

Fig. 2.1 Photograph of as grown LAHSQ single crystal

Table 2.1: Crystal parameters of LAHSQ single crystal

Empirical Formula	$C_6H_{15}N_4O_2 \cdot C_4HO_4$
Formula weight	288.27 g/mol
Crystal system	Triclinic
Space group	P1
Unit cell dimensions	$a = 5.1301(14)$ Å, $\alpha = 93°(5)$ $b = 8.3224(8)$ Å, $\beta = 96.16°$ $c = 14.9005(15)$ Å, $\gamma = 100.17°(5)$
Cell volume	621 Å^3
Absorption coefficient	0.13 mm^{-1}
F (000)	254
Crystal size	0.40 x 0.30 x 0.30 mm^3
Theta range for data collection	3.523 to 22.899°.
Limiting indices	-7<=h<=7, -18<=k<=18, -16<= I <=16
Reflections collected/unique	2143/ 2012
Completeness to theta = 23.92	96.5%
Refinement method	Full-matrix least-squares on F^2
Data/restraints/parameters	2214 / 1 / 176
Goodness-of-fit on F2	1.052
Final R indices [I>2sigma(I)] 80	R1 = 0.0385, wR2 = 0.0789
R indicies (all data)	R1 = 0.0548, wR2 = 0.0847

parameters such as bond length and bond angles of LAHSQ were obtained using the gaussian 09 package programme with (B3LYP)/6-311++G (d, p) basis set. In the LAHSQ structure, there are 8 (C-C) bond with bond length that varies between 1.305-1.542 Å, 9 (C-H) bond length ranging from 1.070-1.254 Å respectively, 8 (N-H) bond with bond length varies from 0.985-1 Å and 1 (O-H) bond having the bond length are at 0.684 Å are observed by (B3LYP)/6-311++G (d, p) basis set. From the table 2.3, the theoretical study revealed that the ring angles of LAHSQ compound are (C1-C4-C10) = 133.700°, (N6-C4-H5) =107.187° and (N18-C20-N21) =120°. The bond angles of less than 120° are caused by hydrogen atoms bound to carbon atoms. The carbon atoms bound to the oxygen atoms have precisely 120° bond length.

2.3.4 VIBRATIONAL ASSIGNMENTS

The vibrational spectrum of LAHSQ have been studied by applying Density Functional Theory with B3LYP/6-311++G (d, p) basis set. The functional groups in the LAHSQ crystal was determined by recording FT-IR spectrum on Bruker IFS 66V spectrometer in the spectral range of 500-4000 cm^{-1} using the KBr pellets and is presented in Fig 2.3. The theoretically computed and experimentally perceived frequencies of Infrared intensities and their various assignments were recorded, and spectra are displayed in Fig. 2.3 & 2.4. The thorough investigation of vibrational assignments of different functional groups is made and details are given here. In experimental the collisional broadening at room temperature in a solvent such as water is significant enough to cause a blurring together of the energy differences between the different rotational and vibrational states, such that the spectrum consists of broad absorption bands instead of discrete lines. It should be noted that the calculations were made for a free molecule in vacuum, while experiments were performed for solid phase or in solution. Furthermore, the anharmonicity is neglected in real system for calculated vibrations, and because of the low IR intensity of some modes, it is difficult to observe them in the IR spectrum. Thus, there are disagreements between calculated and observed vibrational wavenumbers.

VIBRATIONS OF O-H GROUP

The vibrations of stretching mode of O-H group have been observed about 3399 cm^{-1}. The O-H stretching vibrations have been perceived in IR at 3266 cm^{-1} as a strong band. Normally, Hydroxyl in-plane bending mode shows up in the frequency range 1440-1260 cm^{-1}. The IR frequency 1358 cm^{-1} have been allotted to in-plane bending mode of hydroxyl group. The O-H out-plane bending vibration is perceived in a powerful region 700 – 600 cm^{-1}. The intermediate intensity of IR at 616 cm^{-1} is allocated to out plane bending mode of hydroxyl group. The superimposed vibration is appeared at 2620 cm^{-1}.

VIBRATIONS OF AMMONIA

The NH^+ asymmetric stretching vibrational mode occurs at 3177 cm^{-1} and symmetric stretching vibrational mode arises at 3039 cm^{-1}. The bands 3152 cm^{-1} in IR have been attributed to NH^+ asymmetric stretching mode. Generally, the NH asymmetric deformation vibrations

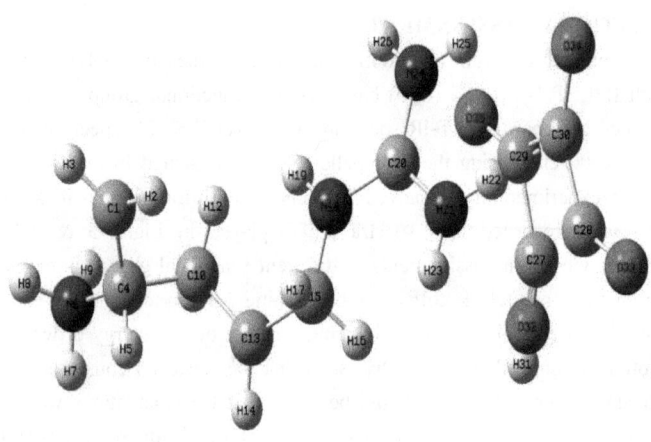

Fig. 2.2 Atomic numbering system adapted for ab initio computations of LAHSQ molecule

Table 2.2: Selected bond lengths of LAHSQ molecule

Parameters	Bond length in (Å)	
	Observed Bond Length	Calculated Bond Length
C1-C4	1.540	1.469
C4-C10	1.458	1.398
C10-C13	1.537	1.502
C10-C15	1.305	1.284
N6-C4	1.470	1.478
C10-H12	1.070	1.289
N6-H8	1	1.112
N6-H9	1	1.021
C10-H11	1.070	1.142
C1-H2	1.070	1.021
N16-C15	1.470	1.392
C20-N21	1.472	1.420
O35-C29	1.430	1.382
O32-C27	1.070	1.011
C28-O33	1.430	1.398
C15-H16	1.150	1.048
C27-H32	1.254	1.396
C29-C30	1.446	1.459
C1-H3	1.070	1.202
C20-N24	1.472	1.736
N21-H23	1.000	1.365
N18-H19	0.985	1.236
H14-C13	1.421	1.489
C15-N18	1.324	1.298
C27-C28	1.542	1.468
O32-H31	0.684	0.592

Table 2.3: Selected bond angle of LAHSQ molecule

Parameters	Bond angles (°)	
	Observed Bond Angle	Calculated Bond Angle
H8-N6-H7	109.471	108.981
N6-C4-H5	107.187	107.547
C10-C13-C15	119.999	120
H17-N18-H19	122.975	121.475
H17-N18-C20	109.335	108.457
N18-C20-N21	120.000	120
H16-C15-H17	108.417	108.214
H31-O32-C27	180.000	179.254
O32-C27-C29	144.189	145.236
O35-C29-C27	125.813	126.256
C29-C29-C30	129.973	130.452
H23-N21-H22	131.781	130.456
C28-C30-O34	136.295	135.246
C1-C4-C10	133.700	108.265
O33-C28-C30	133.100	133.213
H8-N6-H9	109.47	109.12
H5-C4-C10	107.54	106.23
H11-C10-C13	109.47	108.54
C13-H12-C10	109.47	109.24
H19-N18-C15	119.99	119.00
H19-N18-C20	120	119.14
C29-C30-O35	144.22	144.01
C29-C30-O34	125.67	124.51
C15-C13-H14	120	120
H25-N24-H26	119.99	118.20
H11-C10-H12	35.26	34.25
H2-C1-H3	107.43	106.13
N6-C4-H51	107.187	105.124

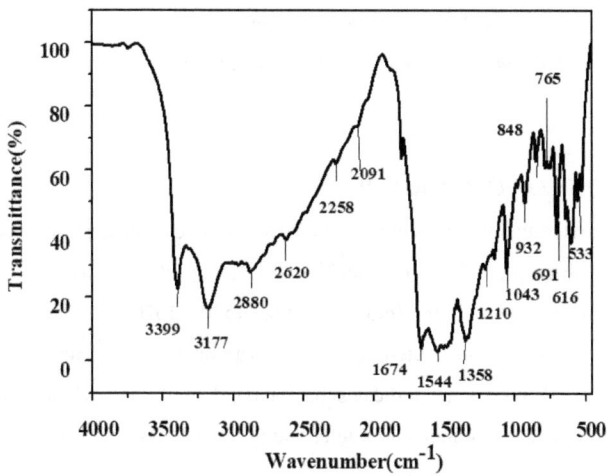

Fig. 2.3 Experimentally obtained FT-IR spectrum of LAHSQ

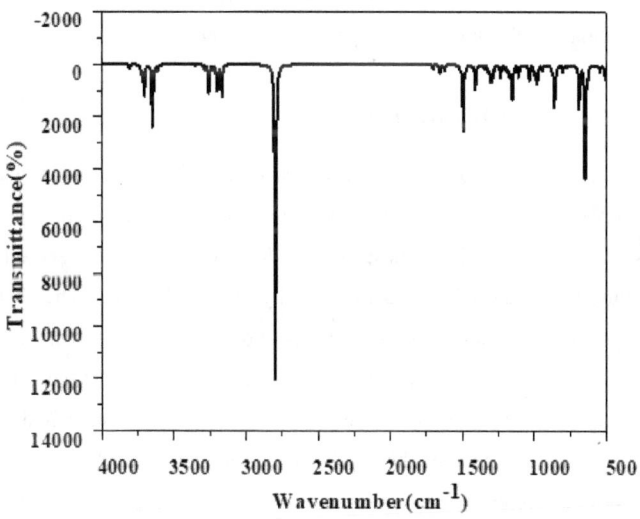

Fig. 2.4 Theoretically simulated FT-IR spectrum of LAHSQ

are identified in the frequency range 1660.1610 cm^{-1} and the symmetric deformations appear in the vibrational range 1550 – 1485 cm^{-1} (Bellamy J. L et. al. 1975). N-H asymmetric deformation is perceived at 1674 cm-1 and N-H symmetric deformation is appeared at 1358 cm$^{-1.}$

VIBRATIONS OF C=O GROUP

The stretching vibrations of C=O create an Infrared characteristic vibrational bands and the structure of hydrogen bond can decide the intensity of these bands. This C=O group of vibrational mode is found in the frequency range 1540 - 1670 cm^{-1}. The active band of IR at 1674 cm^{-1} are allocated to stretching mode of carbonyl group. The C=O stretching mode achieved the red shifting which indicates that the bonding of carbonyl group with other groups. Due to the carbonyl group interaction, inter and intramolecular hydrogen bonding are formed.

VIBRATIONS OF COO⁻ GROUP

The COO- ion brings about two different types of mode. One of those is asymmetric stretching, which is manifested between 1674 and 1544 cm^{-1}. The other is symmetric stretching vibrational mode which is manifested around 1358 cm^{-1}. In LAHSQ, the asymmetric stretching vibrational mode of COO⁻ group is perceived as an active range in IR at 1544 cm^{-1}. The scissoring vibration of COO⁻ group is appeared at 691 cm^{-1}.

2.3.5 HYPERPOLARIZABILITY

The force exerted by the molecular packing and intermolecular interactions in the crystal will play as a decisive factor for the determination of macroscopic properties of these materials. So, understanding these intermolecular interactions should help towards understanding the nature of the macroscopically produced effects. The ab initio calculated non-zero µ value shows that this compound might have microscopic first static hyperpolarizabilities with non-zero values obtained by the numerical second derivative of the electric dipole moment according to the applied field strength. The magnitude of molecular hyperpolarizability, presence of the number of chromophores and the degree of non-centro symmetry are the deciding criteria of the second order susceptibility $\chi^{(2)}$ values in an NLO system. A large value of the first hyperpolarizability is the prerequisite to behave as a good NLO material, and the important parameters influencing β generally are (i) donor– acceptor system, (ii) nature of substituents, (iii) conjugated π system and(iv) the influence of planarity.

As mentioned above, this study is extended to the determination of the electric dipole moment μ_{tot}, the isotropic polarizability α_{tot} and the first hyperpolarizability β_{tot} of the title compound. It is well known that the nonlinear optical response of an isolated molecule in an electric field $E_i(\omega)$ can be presented as a Taylor series expansion of the total dipole moment, μ_{tot}, induced by the field:

$$\beta_{tot} = \mu_0 + \alpha_{ij} E_j + \beta_{ijk} E_j E_k + \ldots$$

where μ_0 the permanent dipole moment, α_{ij} is the linear polarizability and β_{ijk} is the first hyperpolarizability tensor components. The isotropic (or average) linear polarizability is defined as (Zhang et. al. 2004)

$$\beta_{tot} = \frac{\alpha xx + \alpha yy + \alpha zz}{3}$$

First hyperpolarizability is a third rank tensor that can be described by 3×3×3matrix. The 27 components of 3D matrix can be reduced to 10 components due to the Kleinman symmetry (Kleinman D.A.1962) (β_{xyy}, β_{yxy}, β_{yyx}, β_{yyz}, β_{yzy}, β_{zyy}, ... likewise, other permutations also take same value). The output from Gaussian 09 provides 10 components of this matrix as β_{xxx}, β_{xxy}, β_{xyy}, β_{yyy}, β_{xxz}, β_{xyz}, β_{yyz}, β_{xzz}, β_{yzz}, β_{zzz}, respectively. The components of the first hyperpolarizability can be calculated using the following equation (Sun et. al. 2003).

$$\beta_i = \beta_{iii} + \frac{1}{3} \Sigma_{i \neq j} (\beta_{ijj} + \beta_{jij} + \beta_{jji})$$

Using the x, y and z components of β, the magnitude of the first hyperpolarizability tensor can be calculated by:

$$\beta_{tot} = (\beta^2 x + \beta^2 y + \beta^2 z)^{1/2}$$

The complete equation for calculating the magnitude of β from Gaussian 98 output is given as follows:

$$\beta_{tot} = ((\beta_{xxx} + \beta_{xyy} + \beta_{xzz})^2 + (\beta_{yyy} + \beta_{yzz} + \beta_{yxx})^2 + (\beta_{zzz} + \beta_{zxx} + \beta_{zyy})^2)^{1/2}$$

The total static dipole moment μ, the mean polarizability α_0, the anisotropy of the polarizability Δα and the mean first hyperpolarizability β_0, using the x, y, z components they are defined as

$$\mu = \mu_x^2 + \mu_y^2 + \mu_z^2$$

$$\alpha_0 = (\alpha_{xx} + \alpha_{yy} + \alpha_{zz})/3$$

$$\beta_0 = (\beta_x^2 + \beta_y^2 + \beta_z^2)^{1/2}$$

Table 2.4: The electric dipole moment μ, the average polarizability $α_{tot}$ and First order hyperpolarizability $β_{tot}$ for LAHSQ

Dipole Moment in Debye	
$μ_x$	2.7755
$μ_y$	2.6202
$μ_z$	-2.4118
$μ_{tot}$	1.7274
Polarizability in esu	
$α_{xx}$	47.66
$α_{xy}$	2.28
$α_{yy}$	37.3
$α_{xz}$	2.77
$α_{yz}$	1.15
$α_{zz}$	30.72
$α_o$	38.56×10^{-24}
Hyperpolarizability in esu	
$β_{xxx}$	209.6768
$β_{xxy}$	149.0357
$β_{xyy}$	140.7659
$β_{yyy}$	339.1507
$β_{xxz}$	-43.9292
$β_{xyz}$	-71.4309
$β_{yyz}$	-129.4552
$β_{xzz}$	78.8258
$β_{yzz}$	142.4187
$β_{zzz}$	-63.9568
$β_{tot}$	6.90342×10^{-30}

The total molecular dipole moment and mean first hyperpolarizability of LAHSQ is shown in Table 2.4. The connection between the electric dipole moments of an organic molecule having donor – acceptor substituent and first order hyperpolarizability is widely recognized in the literature.

2.3.6 HOMO and LUMO analysis

The highest occupied molecular orbital (HOMO) and lowest unoccupied molecular orbital (LUMO) can assist in determining the nature of intra molecular interaction of LAHSQ. The chemical stability of a molecule can be scaled by frontier molecular orbital energy gap. In general, the HOMO shows the tendency of donating electron and LUMO exhibits electron acceptor character. In the present study the HOMO and LUMO are computed at the rank of B3LYP/6-311++G (d, p) theory and their respective plots are shown in Fig. 2.5 The interaction between HOMO and LUMO is denoted by π-π* (Fukui K.et al 1975) and the energy difference between HOMO and LUMO is called energy gap (Lewis D.F. V. et .al 1994). This energy gap is responsible for the charge transfer process that takes place within the molecule.

The LUMO and HOMO energy levels of LAHSQ are -5.4384 eV and -2.2593 eV, respectively. The HOMO-LUMO energy gap was calculated as 3.1790 eV. The HOMO-LUMO energy, hardness, softness, electronegativity, chemical potential and electrophilicity index of LAHSQ are absorbed by DFT (B3LYP/6-311 ++ G (d, p)) basis with their respective values are lists in Table. 2.5. The chemical hardness (η), Global Softness (S), electro negativity (χ), electronic chemical potential (μ), and electrophilicity (ω) were approximated by DFT analysis are 0.4599, 1.0871, 3.8488, -3.8488 and 16.1048 in the order.

GLOBAL CHEMICAL REACTIVITY PARAMETERS

Chemical harness (η), electronic chemical potential (μ), electronegativity (χ), softness of a molecule (S) and electrophilicity (ω) were estimated using DFT calculations.

Chemical harness was estimated by $\eta = (I-A)/2$

The softness of a molecule was calculated by $S = 1/2\,(\eta)$

Electronegativity was calculated by $\chi = (I+A)/2$

Chemical potential of the molecule was estimated by $\chi = -(I+A)/2$

Electrophilic index was calculated by $\omega = \mu^2/2\,(\eta)$

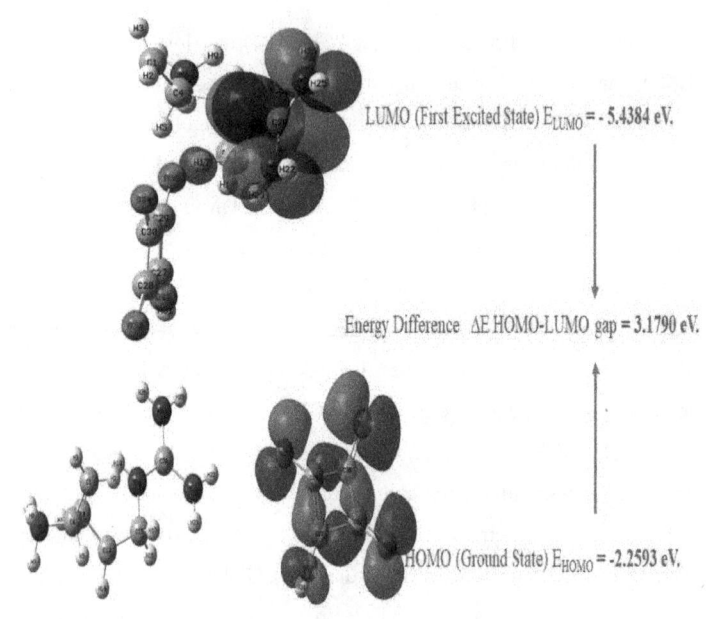

Fig. 2.5 HOMO – LUMO plot of LAHSQ at B3LYP/6-311G++ (d, p)

Table: 2.5: Calculated electronic and energies of LAHSQ using B3LYP/6-311++G (d, p) level

Electronic parameter	Calculated values
E_{HOMO} (eV)	-2.2593
E_{LUMO} (eV)	-5.4384
$\Delta E_{HOMO} - E_{LUMO}$ (eV)	3.1790
Ionization Potential (IP)	2.2593
Electron Affinity (EA)	5.4384
Chemical Hardness (η)	0.4599
Softness (s)	1.0871
Electronegativity (χ)	3.8488
Chemical Potential (μ)	-3.8488
Electrophilicity index (ω)	16.1048

2.3.7 THERMODYNAMICAL PROPERTIES

The standard statistical thermodynamic functions at B3LYP/6311++G (d, p) level heat capacity C_P^0m, entropy S^0m and enthalpy changes ΔH^0m for the title compound are obtained from the theoretical wavenumbers (Table 2.6), which shows that these thermodynamic functions are increasing with temperature ranging from 100 to 1000 K due to the fact that the vibrational band intensities increase with temperature (Ott J.B. et al 2000). The correlation equations between heat capacity, entropy, enthalpy changes and temperatures are listed below and can be used for analysing heat capacities, entropies and enthalpies in different temperatures fitted by quadratic formulas, and the corresponding fitting factors (R^2) for these thermodynamic properties are 0.9972, 0.9994 and 0.99945 respectively. The corresponding fitting equations are as follows and the correlation graphs of these are shown in Fig. 2.6, 2.7 and 2.8.

$$C_P^0m = 31.884+0.446T-2.0773\ T^2\ (R^2=0.9972)$$
$$S^0m = 234.700+0.621T-1.9649\ T^2\ (R^2=0.9994)$$
$$\Delta H^0m = -7.330+0.087T+1.06406\ T^2\ (R^2=0.99945)$$

2.3.8 OPTICAL STUDIES

The spectral examination of an UV-Visible spectrum of the grown LASHQ crystal was carried out using Perkin Elmer LAMBDA-25 UV-Visible spectrometer in the range of 200-1200 nm. The assessment of optical nature of the crystalline material facilitates to categorize its extensive reliability for optoelectronic and NLO device fabrications. Generally, the optical transparency depends on several interior and exterior aspects which consist of the optically dynamic functional units, orientations, grain boundaries, impurities and striations. This factor that exhibits the linear optical transparency in crystal is governed by the electronic transition assist by the interaction of electromagnetic spectrum with crystalline material. The resulted transmission spectrum is displayed in Fig. 2.11.

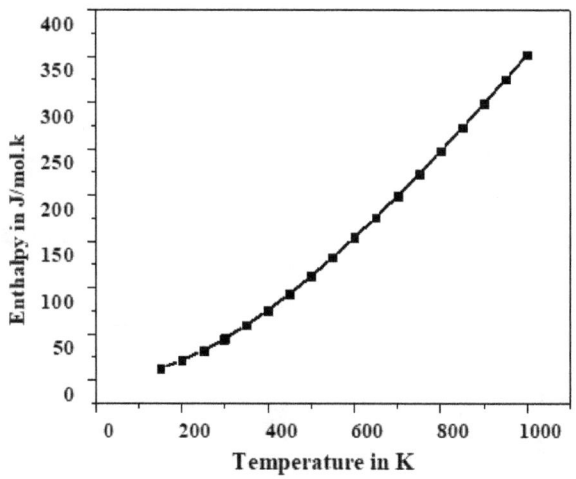

Fig. 2.6 Variation of Enthalpy with temperature

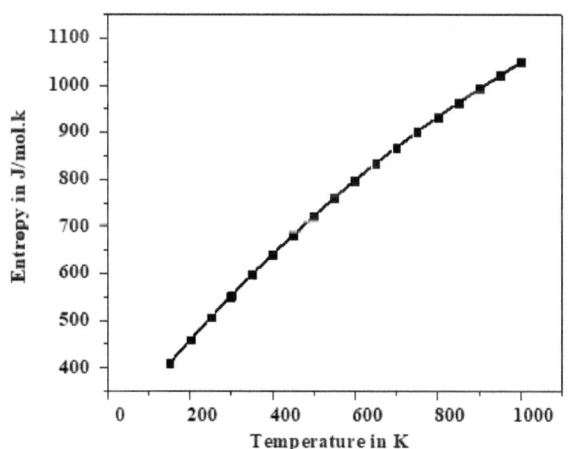

Fig. 2.7 Variation of Entropy with temperature

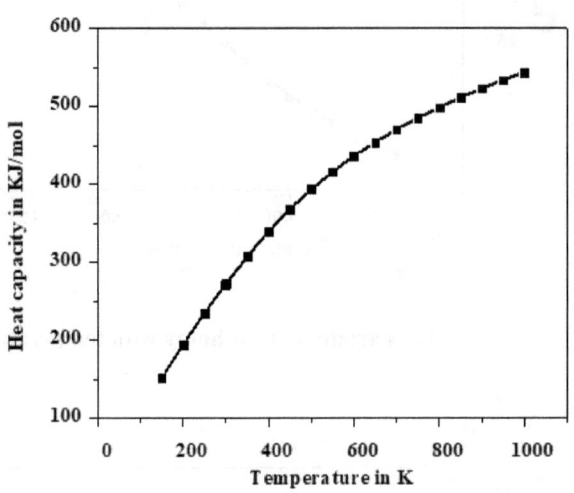

Fig. 2.8 Variation of heat capacity with temperature

Table 2.6 Thermodynamic properties at different temperatures by B3LYP level for LAHSQ

T (K)	S (J/mol.K)	Cp (J/mol.K)	ΔH (kJ/mol)
200	507.003	194.158	21.651
250	551.476	234.491	32.376
298.15	553.157	271.173	44.559
300	597.85	272.533	45.062
350	641.054	307.698	59.581
400	682.724	339.552	75.776
450	722.831	368.01	93.479
500	761.387	393.274	112.524
550	798.432	415.703	132.759
600	834.028	435.704	154.054
650	868.253	453.663	176.296
700	901.186	469.915	199.392
750	932.911	484.733	223.264
800	963.504	498.334	247.845
850	993.039	510.887	273.079
900	1021.583	522.523	298.918
950	1049.2	533.346	325.318
1000	1049.2	543.437	352.241

From UV transmittance spectra, it is known that LAHSQ crystals have high transmittance from 350 –1200 nm. This is very important for materials possessing nonlinear optical properties. The cut of wavelength of grown crystal is observed as 364 nm. The wide range of transparency (90%) is an added advantage for this crystal to be utilized in the field of optoelectronic devices. The optical energy gap was calculated as 3.20 eV is shown in Fig.2.10. The lower cut-off wavelength was observed as nearly 364 nm is shown in Fig. 2.9. The broad range of transmittance is observed throughout the visible spectral region of the optical absorption spectrum.

Extinction coefficient (K) can be derived from absorption coefficient according to equation which indicates the rate of loss of electromagnetic wave through scattering and absorption when electromagnetic wave propagates through a crystal. According to the absorption coefficient and the formula, the reflectance (R) was calculated.

$$K = \alpha\lambda/4\pi$$

$$R = 1 \pm \frac{\sqrt{1 - \exp - (\alpha t) + \exp \alpha\,(t)}}{1 - \exp(\alpha t)}$$

$$n = \frac{-(R+1) \pm \sqrt{3R^2 + 10R - 3}}{2R - 1}$$

Figs. 2.12 to 2.14 shows the variation of extinction coefficient (K), refractive index (n) and reflectance (R) as a function of wavelength for LAHSQ crystal. From the graph, it is also noted that the extinction coefficient (K) increases with increasing of the incident light wavelength. And the variation in reflectance (R) and extinction coefficient may be caused by the interaction between photons and electrons.

2.3.9 MULLIKEN POPULATION ANALYSIS

The vibrational spectra are mainly influenced by the charge distribution among the molecules. The Mulliken population analysis of LAHSQ molecule is calculated by using B3LYP/6-311++G (d, p) basis set, were listed in table 2.7 and are displayed in Fig. 2.15.

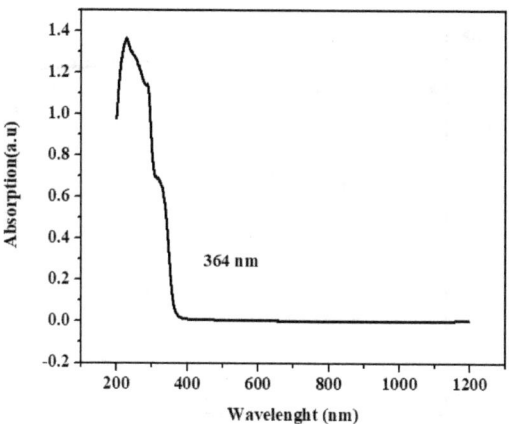

Fig. 2.9 UV- Vis absorption spectrum of LAHSQ crystal

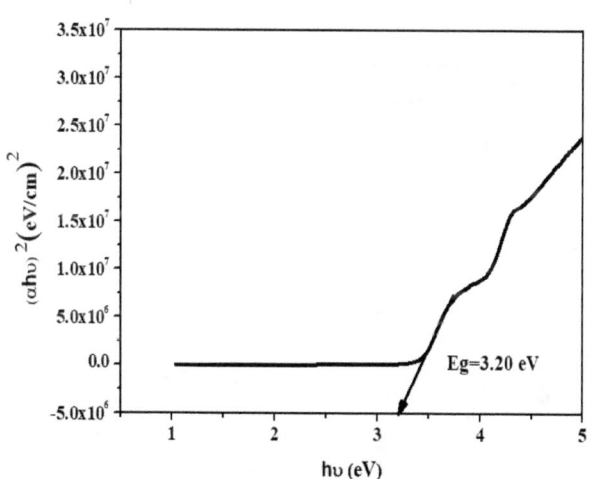

Fig. 2.10 UV- Vis bandgap of LAHSQ crystal

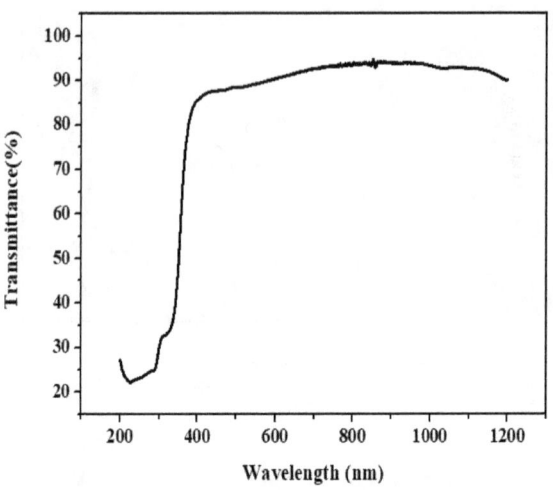

Fig. 2. 11 UV-Vis transmission spectrum of LAHSQ crystal

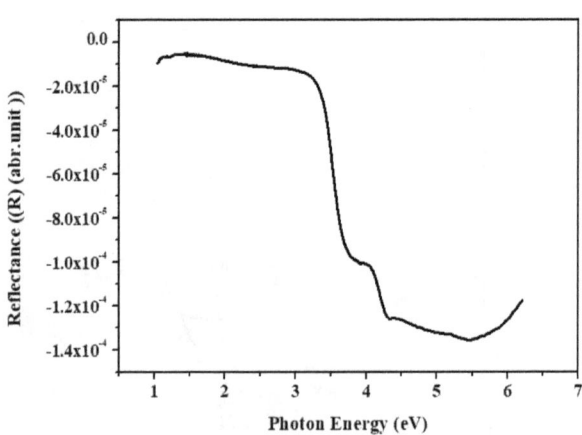

Fig. 2. 12 UV- vis reflective index of LAHSQ crystal

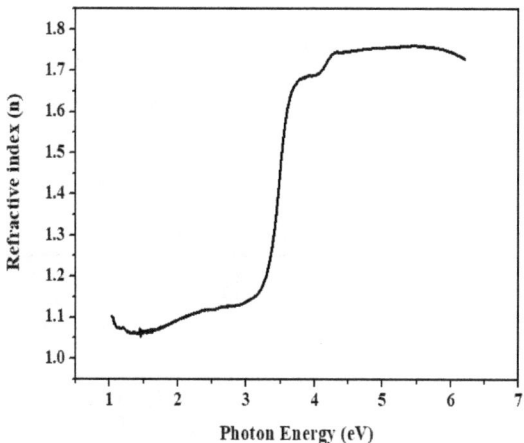

Fig. 2. 13 UV- Vis refractive index of LAHSQ crystal

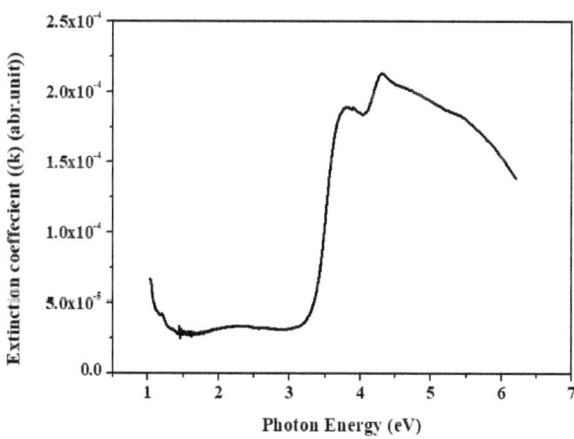

Fig. 2. 14 UV-Vis extinction coefficient of LAHSQ crystal

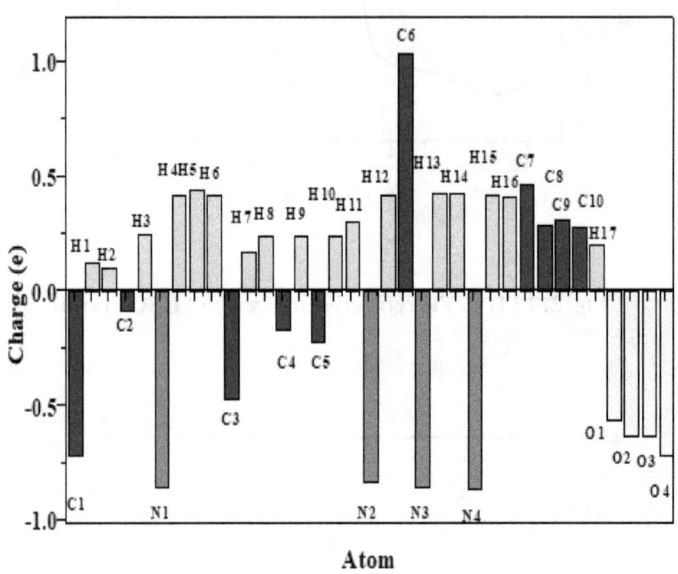

Fig. 2.15 Mulliken atomic charges of LAHSQ single crystal

Table 2.7: Mulliken atomic charges of LAHSQ single crystal

Atoms	Mulliken atomic charge
C1	-0.7152
H1	0.1232
H2	0.0967
C2	-0.0808
H3	0.2472
N1	-0.8657
H4	0.4137
H5	0.4297
H6	0.4137
C3	-0.4735
H7	0.1667
H8	0.2472
C4	0.1671
H9	0.2311
C5	-0.2368
H10	0.2361
H11	0.2953
N2	-0.8386
H12	0.4192
C6	1.0321

N3	-0.8601
H13	0.4192
H14	0.4081
N4	-0.8601
H15	0.4032
H16	0.4081
C7	0.4624
C8	0.2903
C9	0.3064
C10	0.2848
H17	0.2096
O1	-0.5697
O2	-0.6289
O3	-0.6400
O4	-0.7208

Carbon is the topmost positive charge possessed atom (~1.0) among the other positive charge possessed atoms like H1, H2 and H17. The negative charge possessed atoms like N (~0.8), O1, O2, O3, O4 (~0.6-0.7) and C1-C5 molecule possess the negative charge. Due to the electron drop out character, Carbon 6 atom keeps the extreme positive charge.

2.3.10 NATURAL BONDING ORBITAL (NBO) ANALYSIS

For all possible interactions between the donor (i) and acceptor (j) stages, the NBO analysis is acknowledged. The molecular bonds can be reinforced or broken by these intermolecular interactions. Delocalization interactions within i and j were linked to the stabilisation energy ($E^{(2)}$) stage. The fact that they have a high ($E^{(2)}$) value indicates that they have good interactions within the door (i) and acceptor (j). in a base range of DFT/B3LYP6-311++ G (d, p), the LAHSQ compound was analysed using NBO analysis. As a consequence, to describe conjugative interactions, intermolecular rehybridization and electron density delocalization within the molecule (Szafran M et.al. 2007). The Fock matrix was used to evaluate the NBO basis using second order perturbation theory. The strong hyperconjugative intermolecular interaction implied by the molecule is shown in table 2.8.

π^* (C13-O19) and n1 (O10) have some intermolecular interactions between antibonding with LP and stabilization energy 20.71 Kcal/mol, that represents large delocalization. Anthor intermolecular interaction is related to orbital overlapping the π^*(C13-O19) from the orbital of π^* (C12-O18), which has high stabilization energy of 260.93 Kcal/mol, indicating large delocalization. σ (O20) and n1 (C14) interaction with antibonding to LP showed a maximum stabilization energy of 160.66 Kcal/mol. Transitions from π to π^* occur between the molecules and have a high polarisation, indicating that the LAHSQ complex has the most NLO activity.

2.3.11 MICROHARDNESS TEST

The Hardness of the crystal carries information about molecular binding, yield strength and elastic constants of the material. The Mechanical properties of the grown crystals have been studied using microhardness tester fitted with a Vicker's diamond pyramidal indenter. The grown crystals which were well polished have been placed on the platform of Vickers microhardness tester. Loads of different magnitudes have been applied in a fixed interval of time. Vickers microhardness values have been calculated by using the formula

$$H_v = 1.8544 \times P/d^2 \text{ Kg/mm}^{-2}$$

Table 2.8: Second order perturbation theory analysis of Fock Matrix in NBO for LAHSQ with 6-311++G (d, p) basis set

Donor (i)	Type	Acceptor (i)	Type	E(2)[a] (k cal/mol)	E(j)-E(i)[b] (a.u)	F(I,j)[c] (a.u)
O9	π*	C5-C8	π*	10.35	0.53	0.068
O9	π	C9-C8	σ*	21.53	0.44	0.087
O10	π	C8-C9	σ*	39.00	0.35	0.108
C8-O10	σ*	N1-C5	σ*	16.46	0.01	0.051
O10	n1	C13-O19	π*	20.71	0.39	0.085
C13-O19	π	C14	n1	23.78	0.13	0.063
C13-O19	π	C12-O18	π*	11.29	0.25	0.051
C14-O13	σ	H16-O20	σ*	34.07	2.93	0.282
C15-O17	π	C14	n1	20.06	0.12	0.058
C15-O17	π	C12-O18	π*	10.87	0.24	0.051
H16-O20	σ	C14-O20	σ*	16.38	2.14	0.168
O20	n1	C14	π*	10.49	19.64	0.405
C14	n1	C13-O19	π*	95.28	0.10	0.105
C14	n1	C15-O17	π*	100.52	0.12	0.116
O17	σ	C12-C15	σ*	10.82	0.69	0.077
O17	n1	C14-C15	σ*	12.53	0.68	0.082
O18	n1	C12-C13	σ*	13.43	0.59	0.080
O18	n1	C12-C15	σ*	10.22	0.70	0.076
O19	n1	C12-C13	σ*	11.56	0.62	0.075
O19	π	C13-C14	σ*	29.00	0.85	0.140
O20	σ	C13-C14	σ*	26.24	0.91	0.138
O20	σ	C14	n1	160.66	0.27	0.220
C13-C14	σ*	C12-C15	σ*	26.03	0.07	0.127
C13-O19	π*	C12-O18	π*	260.93	0.02	0.085

$E^{(2)}$ means energy of hyperconjugative inetractions
Energy difference between donor and acceptor i and j NBO orbital
F(i,j) is the Fock matrix element between i and j NBO orbital .
n1 – lone pair

where H_v is the Vickers microhardness number, P is the applied load in Kg, d is the mean diagonal length of the indentation impression in mm and 1.8544 is a constant of a geometrical fraction for the diamond pyramid. The trace of Vickers hardness number with load for LAHSQ crystals are shown in the Fig. 2.16 this shows that the hardness increased with the increase of load. The hardness of the material is guaranteed from the plot between log P and log d in Fig. 2.17. Higher the hardness values, greater the stress required to form dislocation, thus confirming greater crystalline perfection. In the present study, n is found to be greater than 1.6, thus confirming that LAHSQ is a soft material.

2.3.12 THERMAL ANALYSIS

Thermogravimetric and differential thermogravimetric (TG/DTA) studies on LAHSQ were carried out at a heating rate of 20°C/min in nitrogen atmosphere at the temperature of range 30–400°C. Fig. 2.18. Perkin-Elmer Diamond analyzes was used to analyze TG-DTA in the nitrogen atmosphere at a heating rate of $10°C\ min^{-1}$. A thin and sharp endothermic peak occurs in the DTA curve at 398 °C, which is near the melting point of the material. The first stage of weight loss occurred at 398°C with 31% volatile chemicals discharged. The TG curve explains the decomposition nature of the material and it was absorbed at the temperature range of 398.54°C. Hence, the LAHSQ material is stable up to 398.54 °C and it was suitable for the device fabrications.

2.3.13 DIELECTRIC STUDIES

The selected grown crystals were cut and polished to obtain about 2 mm thickness. It was subjected to dielectric studied at room temperature using HIOKI 3532-50 LCR HITESTER meter in the frequency region 50 Hz -5 MHz. The dielectric analysis is an important characteristic than can be used to fetch the knowledge about the electrical properties the material as a function of frequency. Dielectric properties are correlated with the electro optic property of the crystal particularly when they are non-conducting material (Hemaraju B.C and Gana Prakash P 2015). The variations of dielectric parameters of the samples with frequencies are shown in the Fig. 2.19 and 2.20. From the graphs, it is observed that dielectric parameters like dielectric constants and loss factors decrease with increase in frequency and the high values of ε_r at low frequencies may be due to presence of space charge polarization and its low value at high frequencies may be due to the loss of significance of the four type of polarizations viz.

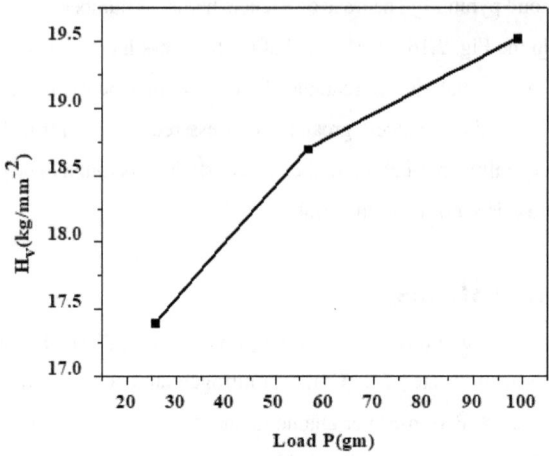

Fig. 2.16 Vickers hardness number Vs applied load of LAHSQ

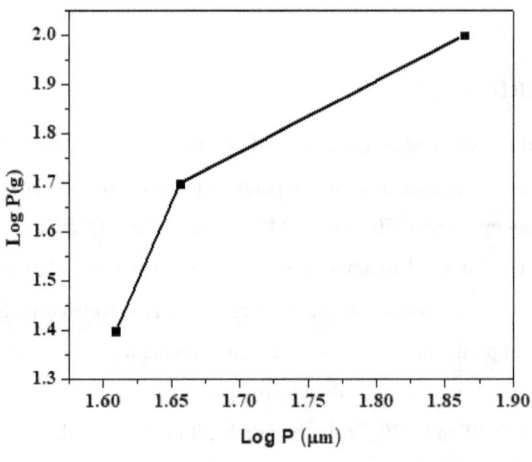

Fig. 2.17 log P Vs log d plot of LAHSQ

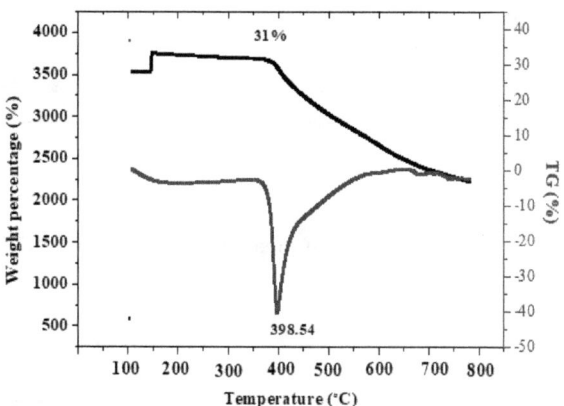

Fig. 2.18 TG-DTA plot of LAHSQ crystal

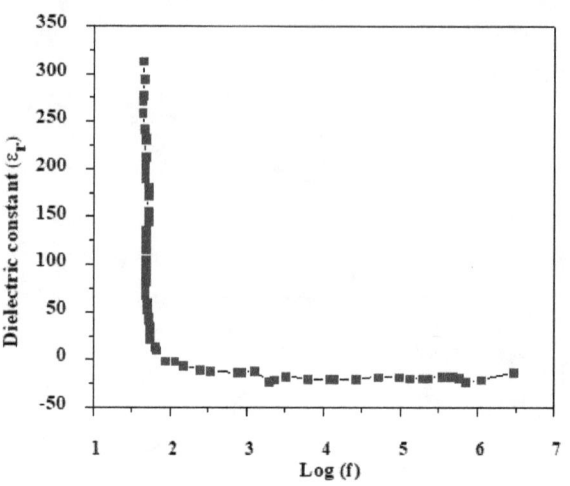

Fig. 2.19 Variation of dielectric constant of LAHSQ

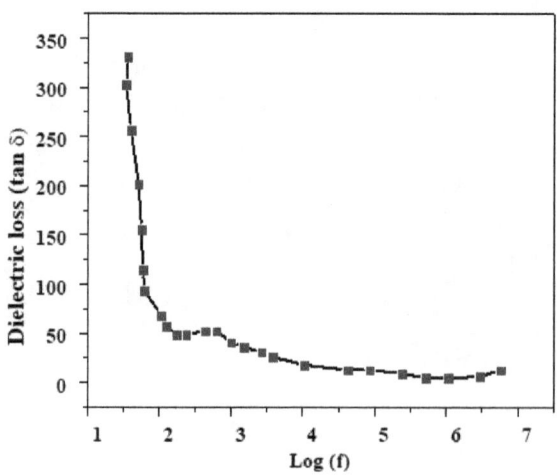

Fig. 2.20 Variation of dielectric loss of LAHSQ

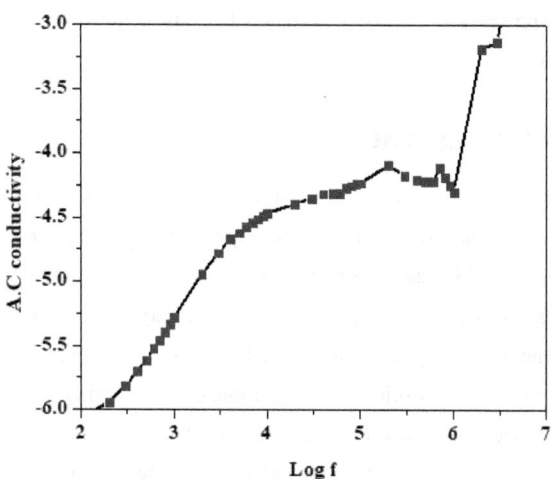

Fig. 2.21 Frequency dependence of AC Conductivity

space charge, orientational, ionic and electronic polarization. In accordance with Miller's rule, the low value of dielectric constant at higher frequencies may be due to the fact that the dipoles cannot follow up the fast variation of the applied field and is a suitable parameter for the enhancement of SHG coefficient and extending the samples application towards photonic, electro-optic and NLO devices. From Fig. 2.21., it can be clearly seen that the value of σ_{ac} is very low up to 6 MHz and increase with increase of frequency and can be explained on the basis of frequency power law. This is the common dielectric behaviour and it refers that it used for NLO applications.

2.3.14 PHOTOLUMINESCENCE ANALYSIS

Photoluminescence spectrum arises when electronic states are excited by certain wavelengths, leading to the emission of photons (Jyoti Dalal and Binay Kumar 2016). Using a PERKIN ELMERLS 45 spectro-fluro photometer and xenon light as the action source (150 W), the intrinsic lead in the band region of the crystal is demonstrated using the luminescence phenomenon. Fig. 2.22 shows the PL graph of the LAHSQ single crystal. The emission of photons due to the excitation of electronic states by certain wavelength give rise to photoluminescence spectrum (Jyoti Dalal et al 2016). Photoluminescence studies are able to identify the luminescence property, recombination mechanisms, deeper defects and dislocations present in the grown crystal. Aromatic organic compounds contain localized π-electrons which are the main objective of the luminescence property and they are highly stable in nature and exhibit fluorescence. The emission peak at 490 nm contributes to blue light emission which denotes LAHSQ may be used as blue light emission source in LED's applications.

2.3.15 MOLECULAR ELECTROSTATIC POTENTIAL

The molecular electrostatic potential (MEP) is related to the electronic density. The colour grading of resulting surface instantaneously shows the molecular size, shape and electrostatic potential value which are very useful in research of molecular structure with its physiochemical property relationship. MEP for the LAHSQ molecule is calculated by B3LYP level with the 6–311++G (d, p) method as shown in Fig. 2.23. The different values of the electrostatic potential at the surface are represented by different colours. Colour ranges from red to blue signifying the electron rich site (Nucleophilic region) to electron deficient site (Electrophilic region) (Okulik, Jubert et. al. 2005). Obviously, electron rich site (Reddish) found on more electronegative atoms location like O32 to O35 and the electrophilic site

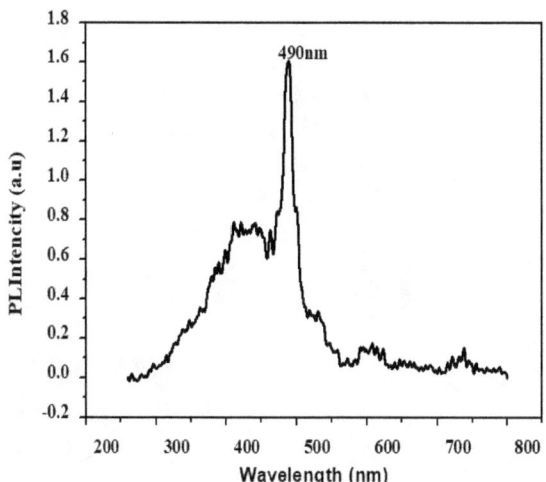

Fig. 2.22. Fluorescence emission spectra of LAHSQ crystal

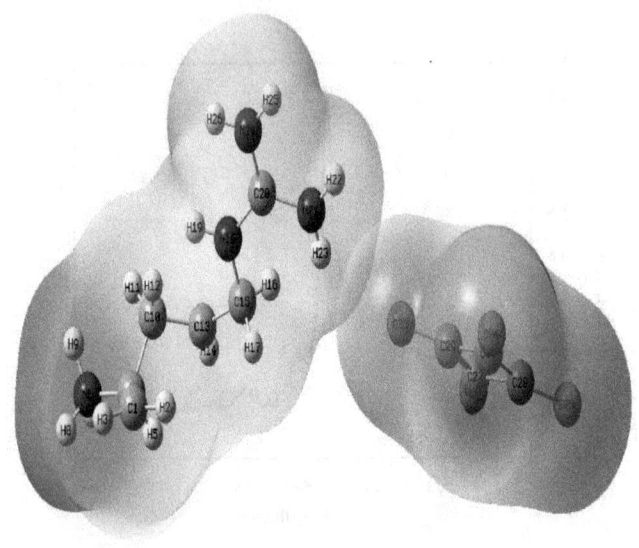

Fig. 2.23. Molecular electrostatic potential (MEP) of LAHSQ crystal

(Blueish) resides at H19, H22, H23, H25, H26 atoms due to the attachment of three electronegative atoms, N18, N21 and N24. The total self-consistent field electron density ranges from -0.150 × e^0 and 0.150 × e^0. MEP pattern was supported by partial atomic charge initiation as well. The MEP provides a visual representation of the chemically active sites and comparative reactivity of the atoms.

2.3.16 Second harmonic generation (SHG)

The Kurtz-Perry (Kurtz et al 1968) powder technique has been performed to investigate the frequency doubling property of LAHSQ crystal. A microcapillary tube was packed by the powdered sample and has been lightened by Q-switched Nd:YAG laser which has the output of 1064nm. The input power to the laser source is 1.2 mJ/pulse. The second harmonic radiation (green radiation) was obtained at 532 nm when the fundamental beam (1064 nm) was sieved by an IR filter. Photomultiplier has used to detect the SHG output and it was displayed by an oscilloscope. The title crystal is 5.5 times greater in second order efficiency (110 mV) when compared with the standard KDP (20 mV). The LAHSQ single crystal is 5.5 times greater in second order efficiency, when compared with the Urea as the reference material (61 mV). From this study to conclude that the LAHSQ crystal would be an optional source in SHG device fabrications.

2.4 CONCLUSION

Organic LAHSQ was cultivated using the slow evaporation method. FT-IR was used to identify the functional groupings. The structural property of LAHSQ was studied by Single crystal X-ray diffraction technique. The single crystal X-ray diffraction data shows that the grown LAHSQ crystal belongs to the P1 space group with triclinic crystal system. The theoretical results showed overall agreement with the experimental record, but they also exhibited significant discrepancies. The entire visible area shows a low absorption in the UV–vis spectrum.

Furthermore, reflectance (R), extinction coefficient (k), and refractive index (n) were calculated. The molecular energy gap is uncovered via the research of border molecular orbitals. The NBO and mulliken molecule charges were determined and explained. The Meyer's analysis observed that the LAHSQ is belongs to a soft material. The thermally stable range of LAHSQ is 398°C. Dielectric investigations demonstrate that both the dielectric constant and loss of the sample diminish as the frequency is increased. In the PL spectrum, the emission peak at 490 nm contributes to blue light emission which denotes LAHSQ may be used as blue light emission source in LED's applications. The efficiency of SHG in grown crystals is 5.5 times that of reference KDP.

CHAPTER III

EXPERIMENTAL AND COMPUTATIONAL STUDIES ON L- GLYCINIUM HYDROGEN SQUARATE (LGHSQ) SINGLE CRYSTAL

3.1 INTRODUCTION

In past few decades, abundant researchers have been focused on the development of new non-linear optical (NLO) materials because of their vital applications in the fields such as generation of higher harmonic frequencies, frequency mixing, telecommunication, electro-optic modulation, optical disk data storage device, optical parametric oscillator, optical disk data storage device (Pecaut J et. al. 1993). etc, Basically, the nonlinear optical materials (NLO) are expected to own enormous nonlinear optical coefficients, suitable transparency and phase matchable properties. In Organic molecules, owing to their molecular flexibility it also exhibits to improve the NLO properties in an effective manner than the inorganic molecules. Organic single crystal possesses the potential advantages than the inorganic materials in NLO applications, including higher hyperpolarizability (Mashraqui S.H et al 2004). Nonlinear optical (NLO) materials, which can generate second harmonic blue-violet light, are of great interest for various applications including optical computing, optical communication, optical disk data storage, optical information processing laser fusion reactions, optical information processing, laser remote sensing, colour display, medical diagnostics etc. Materials with large second-order optical nonlinearities, lower-cut-off wavelength and stable physicochemical performance are needed in order to realize many of photonic and optoelectronic applications. Considerable efforts are currently made to develop new organic materials with large second-order nonlinear optical (NLO) susceptibilities (Patil et. al. 2007). It has been generally that the second-order molecular nonlinearity can be enhanced by large delocalized π-electron systems with strong donor and acceptor groups. The significance of amino acids in NLO applications is due to the fact that all have chiral symmetry and crystallizes in non-centrosymmetric space groups (Senthil S et. al. 2009). Of all amino acids Glycine has been used most frequently for the synthesis of salts of amino acids and neutral compounds. Glycine is well soluble in water and tends to incorporate readily into crystals. Some amino acids like glycine and L-arginine by itself have higher SHG conversion efficiency. The addition of amino acids of glycine exists as Zwitterionic nature favouring crystal hardness. This chapter deals with the experimental and theoretical investigation of L- Glycinium Hydrogen Squarate (LGHSQ) single crystal.

The present work deals with the growth and detailed vibrational spectral investigation of LGHSQ to explicate the correlation between the molecular structure and NLO property, charge transfer interaction between HOMO-LUMO, bond length and first hyperpolarizability aided by using B3LYP levels with 6-311++G (d, p) technique based on DFT computation. The grown crystals have been subject to different characteristics such as single crystal XRD, FTIR, UV-Vis, PL, TG/DTA, dielectric study, Microhardness study, Photoluminescence studies and NLO application of LGHSQ has been reported and the obtained results are elaborately discussed.

3.2 SYNTHESIS AND GROWTH

L-Glycine – $C_2H_5NO_2$ (Sigma–Aldrich, 99%) and Squaric acid ($C_4H_2O_4$) (Sigma–Aldrich, 99%) were taken in equimolar ratio (1:1). The L-Glycinium hydrogen Squarate (LGHSQ) single crystals has been grown by the method of slow evaporation process at room temperature. The solution was mixed together until to attain the supersaturated level and the solution was filtered, decanted into a 150 mL beaker and sealed with silver foil paper, then the beaker was housed for slow evaporation method in a dust free atmosphere. The LGHSQ single crystal was harvested within 40-45 days from the mother solution. The photograph of LGHSQ single crystal is shown in Fig. 3.1.

3.3 RESULTS AND DISCUSSION

3.3.1 SINGLE CRYSTAL X-RAY DIFFRACTION

In order to determine the cell parameters and space group of the grown crystal was confirmed by single crystal XRD. The cell parameters of LGHSQ were found that a = 7.265Å, b=21.900Å, c=10.017Å, α=90°, β=95.43°, γ=90° and V=1586Å^3 and the crystallographic data and refinement details are given in Table 3.1. From the X-ray diffraction data, it is known that LGHSQ crystallises in Monoclinic crystal system with P2$_1$ space group which is non-centrosymmetric, thus fulfilling one of the basic and important material requirements for SHG activity of the crystal.

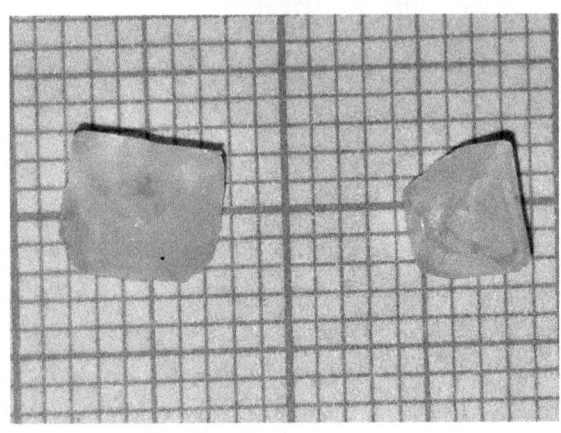

Fig. 3.1 Photograph of as grown LGHSQ single crystal

Table 3.1: Crystal parameters of LGHSQ single crystal

Empirical Formula	$C_6 H_7 N O_6$	
Formula weight	189.13 g/mol	
Crystal system	Monoclinic	
Space group	$P2_1$	
Unit cell dimensions	a = 7.265Å	α = 90°
	b = 21.900Å	β = 95°
	c = 10.017Å	γ = 90°
Cell volume	1586.65 Å^3	
Absorption coefficient	0.145 mm^{-1}	
F (000)	784	
Crystal size	0.300 x 0.250 x 0.200 mm^3	
Theta range for data collection	2.763 to 24.999°.	
Limiting indices	-8<=h<=8, -26<=k<=26, -11<=l<=11	
Reflections collected/unique	49291	
Completeness to theta = = 24.999°	99.0 %	
Refinement method	Full-matrix least-squares on F^2	
Data/restraints/parameters	2764 / 0 / 276	
Goodness- of – fit on F^2	1.178	
Final R indices [I>2sigma(I)] 80	R1 = 0.0656, wR2 = 0.2093	
R indicies (all data)	R1 = 0.0714, wR2 = 0.2144	

3.3.2 COMPUTATIONAL DETAILS

The optimized structural characteristics and the normal vibrational modes of L-Glycinium Hydrogen Squarate (LGHSQ) molecule were studied by using Density Functional Theory (DFT) calculations. The original molecular geometry of the L-Glycinium Hydrogen Squarate (LGHSQ) single crystal was taken from the Crystallographic Information File (CIF) obtained from the single crystal XRD structure analysis. Based on this optimized structure, the vibrational frequencies of the molecules and their intensities also calculated. All the calculated vibrational modes of the molecules are positive and this report shows that the structures are stable.

In addition, the HOMO-LUMO energy gap of the molecules, nonlinear optical properties of the materials, and other physical properties of the LGHSQ single crystal were investigated by using the optimized structures and the vibrational modes of the compound. Molecular Electrostatic Potential (MEP) can be identifying the electrophilic and nucleophilic regions of the molecules within the crystal.

3.3.3 MOLECULAR GEOMETRY

The optimized structural parameters of L-Glycinium Hydrogen Squarate (LGHSQ) single crystal were calculated B3LYP levels with the 6–311++G (d, p) basis set and the values are recorded in Table. 3.2. The calculated structure of the molecule and the atom-labelling scheme adopted for the calculation are shown in Fig.3.2. The LGHSQ molecule has five C-C bonds, one C-N bonds, two C-H bonds, two H-O, three N-H bonds and six C-O bond, one O-O bond. The enormously close similarity has been observed between the geometrically optimized structure of the LGHSQ compound and the actual crystal structure of this molecule. DFT calculations, using B3LYP/6–311++G (d, p) basis sets, produced bond lengths of 1.255 Å and 1.460 Å for the C5-C8 and C8-O9 bonds, respectively. The carbon atoms in the molecule are bound to oxygen atoms that form a square shape. The sp^3 carbon atom has a magnetic field imposed on its electron cloud as a result of squaric acid substitution. This ring of oxygen atoms increases the C-H force constant, which in turn causes bond length to decrease. Due to the repulsion between the lone pair of an electron on the oxygen atom and an electron on the carbon atom in the ring, the C13-O19 bond length is longer (1.503 Å). In the single crystal of LGHSQ, the O10-C8-C5 bond angle is 111.139°, while the C12-C15-O17 bond angle is 136.614°. Valence shell electron pair repulsion theory was used to explain the variation of bond angles. As indicated in Table 3.2 and 3.3 the values of the selected bond

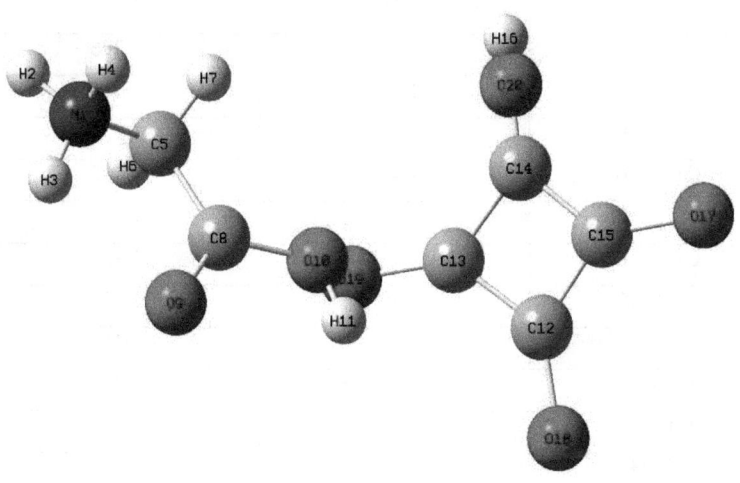

Fig. 3.2 Atomic numbering system adapted for ab initio computations of LGHSQ molecule

Table 3.2: Selected bond lengths of LGHSQ molecule

Parameters	Bond Length (Å)	
	Observed Bond Length	Calculated Bond Length
C5 – C8	1.255	1.321
C12 – C13	1.441	1.396
C13 – C14	1.480	1.423
C12 – C15	1.316	1.296
C14 -C15	1.465	1.411
C14 – O20	1.238	1.201
C13 – O19	1.503	1.492
C12 - 18	1.251	1.232
C15 – O21	1.317	1.302
N1-C5	1.437	1.421
C8 - 09	1.460	1.452
C8 – O10	1.254	1.240
O10-H11	0.976	0.952
N1-H2	1.025	1.246
C5- H6	1.498	1.435
O10-H11	1.093	0.954
C8-O10	1.344	1.321
C5-H7	1.086	0.998
H4-N1	1.022	1.112
C-N1	1.493	1.365
O19-C13	1.241	1.111
C13-12	1.502	1.492
C14-O20	1.335	1.254
O20-H16	0.970	0.965
C15-O17	1.234	1.112
C5-H5	1.093	0.965

Table 3.3: Selected bond angles of LGHSQ molecule

Parameters	Bond Angle (°)	
	Observed Bond Angle	Calculated Bond Angle
O9-C8-O10	124.631	123.124
H11-O10-C8	111.447	110.210
O10-C8-C5	111.139	109.254
H7-C5-H6	110.354	109.235
C5-N1-H4	113.450	111.965
H3-H4-N1	36.508	35.652
H4-N1-H2	109.781	108.541
O19-C12-C13	134.827	133.524
C12-C13-C14	88.937	88.235
C13-C14-C15	94.393	95.654
C14-O20-H16	108.133	107.541
O20-C14-C15	134.536	135.254
C14-C15-O17	135.442	135.214
C12-C15-O17	136.614	135.233
C15-C12-O18	135.782	134.252
N1-C5-H5	135.782	134.632
C8-C5-H7	111.683	110.214
C5-N1-H2	109.560	108.562
O18-C12-C13	135.484	135.214

length and bond angle. The experimentally obtained value was approximately equal to it. The Variation of the bond angles has been described on the basis of Valence shell electron pair repulsion theory. The values of selected bond length and bond length as shown in Table 3.2. and 3.3. It was nearly coincided with the experimental value.

3.3.4 VIBRATIONAL ASSIGNMENTS

The enrolled frequency of FT-IR spectral assignments was achieved based on the theoretically predicted wavenumbers by density functional theory (DFT). The vibrational spectra of LGHSQ single crystal have been studied by applying Density Functional Theory (DFT) with B3LYP/6-311++G (d, p) basis set using gauss view respectively. The experimentally perceived and theoretically computed frequencies of IR intensities and their various vibrational assignments of LGHSQ single crystal and spectral assignment graph are displayed in Fig.3.3. & 3.4. The thorough investigation of vibrational assignments of different functional groups were identified in the LGHSQ single crystal and detailed explanations are given here.

O-H VIBRATION

The stretching vibrational modes of O-H group have been detected at the value of 2982 cm^{-1} (Silverstein R.M et. al. 2005). The presence of O-H stretching vibrational mode occurs in the frequency range of 3200 cm^{-1} and it was considered as a strong band. In General, the frequency of 1440-1214cm^{-1} (Colthup N.B et. al. 1990), has been consider as the Hydroxyl in-plane bending mode of vibration. The in-plane bending mode of hydroxyl group occur in the frequency range of 1372 cm^{-1}. The presence of O-H out-plane bending vibrational mode has been allotted in a powerful region it was observed at range of 700-600 cm^{-1} (Socrates G et. al. 2001). The out-plane bending vibrational mode of hydroxyl group are well agreed with the corresponding wavenumber of 609 cm^{-1} respectively. The superimposed FT-IR vibration has been perceived at nearly 2713 cm^{-1} in the LGHSQ compound.

N-H VIBRATION

The heterocyclic compounds have been perceived in the frequency range of 3500-3000 cm^{-1} (Bellamy L.J et. al. 1975). These values have been considered as N-H stretching vibrational mode. A weaker although sharper band than that of the alcohol (O-H) which is also present in the same region is detected. The NH_2^+ in-plane deformation has been observed at 1639 cm^{-1}. Generally, the N –H asymmetric deformation vibrations were perceived in the frequency range of 1760-1610 cm^{-1} and the symmetric deformations will be appeared in the vibrational range of 1560-1485 cm-1 (C. Karabacak Karaca et al 2012). N-H symmetric deformation is appeared at the frequency range of 1557 cm^{-1} in IR.

C-C VIBRATION

The stretching vibrational mode of (C-C bond) are expected within the region of 1350-1000 cm^{-1}. In the present study, the bands which are of different intensities were observed at 1105, 1121 and 1214 cm^{-1} in the FT-IR spectrum. The presence of aromatic compound and hetero aromatic compounds are allocated to the carbon modes of vibrations and it was perceived at the frequency range between 1400 and 1650 cm^{-1} (H. Saral et al 2016). All the C-C band are occurring within the expected range, which indicates that the C-C vibrations are not altered by the presence of the substitution groups. The in-plane vibration of C-C bond is predicted below the range of 1000 cm^{-1}. The in-plane vibrations are found at 1105, 887, 609 and 754 cm^{-1}. The observed vibrations are in the expected region and they are not affected by any other vibration modes.

C-H VIBRATION

In the presence of aromatic compounds, C-H stretching vibrational modes will be seemed in the range between 3000-3100 cm^{-1}. The C-H bending and stretching vibrational regions are of the most difficult vibrational regions to interpret in infrared (IR) spectra. The IR band at 2757 cm^{-1} has been consider as C-H asymmetric stretching vibration is perceived as a weak IR band. The vibrations of CH_2-twist are observed in the frequency range of 1214-1021 cm^{-1} and CH_2-rock vibrations has been existed in the frequency region of 1160-887cm-1. CH_2 twisting mode is existed in infrared spectra at 1105 cm^{-1} as a poor band and CH_2 rocking vibrational mode is existed in the range of 1021 cm^{-1}, CH_2 wag vibrational mode is supposed to extend over a broad band frequency in the range of (1372 – 1214 cm^{-1}) (H. Tanak et al 2008). CH in-plane bending, and CH out-plane bending are perceived between the frequency range of 1300-1000 cm^{-1} and 887 – 609 cm^{-1} respectively (R.M. Silverstein et al 2005).

C-N VIBRATION

The mixing of several modes is possible in the region makes the documentation of C-N bonds very difficult. The frequency range between 1300 cm^{-1}-1500 cm^{-1} indicates the presence of C-N bonds in the LGHSQ compounds (I. Hubert Joe et al 2009). The vibration bands of the LGHSQ compound has been perceived at the values of 1080, 1056, 966 and 881 cm^{-1} in the FT-IR region. Correspondingly, the in plane and out-of-plane bending vibrational mode are occurred below the frequency range of 491 cm^{-1} and 752 cm^{-1}. The C-N bond was perceived at 1080, 1056, 966 and 881 cm^{-1} in the FT-IR region.

C-O VIBRATION

The carbonyl group expressed the strong absorption band for C-O stretching vibrations. This vibration has been perceived region of 1850-1550 cm^{-1} (M. Saravanan et al 2016). The recorded spectrum shows a sufficient strong band in FT-IR spectrum at the values of 1202 cm^{-1} and at 824 cm^{-1} in FT-Raman spectrum respectively.

VIBRATIONS OF O-H GROUP

The vibrations of stretching mode of O-H group have been observed about 3399 cm^{-1}. The O-H stretching vibrations have been perceived in IR at 3266 cm^{-1} as a strong band. Normally, Hydroxyl in-plane bending mode shows up in the frequency range 1440 –1260 cm^{-1}. The IR frequency 1358 cm^{-1} have been allotted to in-plane bending mode of hydroxyl group. The O-H out-plane bending vibration is perceived in a powerful region 700 – 600 cm^{-1}. The intermediate intensity of IR at 616 cm^{-1} is allocated to out plane bending mode of hydroxyl group. The superimposed vibration is appeared at 2620 cm^{-1}.

VIBRATIONS OF AMMONIA

The NH^{+} asymmetric stretching vibrational mode occurs at 3177 cm^{-1} and symmetric stretching vibrational mode arises at 3039 cm^{-1}. The bands 3152 cm^{-1} in IR have been attributed to NH^{+} asymmetric stretching mode. Generally, the NH asymmetric deformation vibrations are identified in the frequency range 1660 – 1610 cm^{-1} and the symmetric deformations appear in the vibrational range 1550 – 1485 cm^{-1} (Bellamy et al 1975). N-H asymmetric deformation is perceived at 1674 cm-1 and N-H symmetric deformation is appeared at 1358 cm$^{-1.}$

VIBRATIONS OF C=O GROUP

The stretching vibrations of C=O create an Infrared and Raman characteristic vibrational bands and the structure of hydrogen bond can decide the intensity of these bands. This C=O group of vibrational mode is found in the frequency range 1540-1670 cm^{-1}. The active band of IR at 1674 cm^{-1} are allocated to stretching mode of carbonyl group. The C=O stretching mode achieved the red shifting which indicates that the bonding of carbonyl group

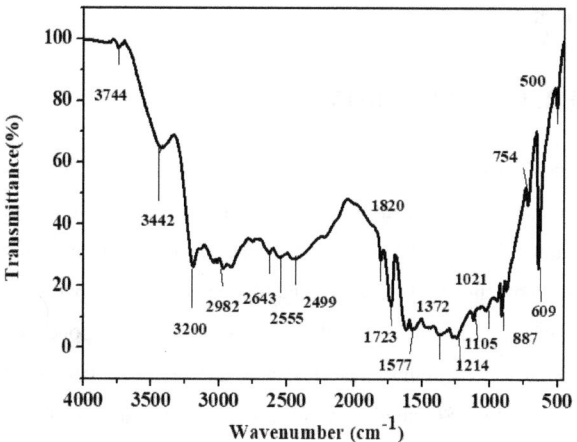

Fig. 3.3 Experimentally obtained FT-IR spectrum of LGHSQ

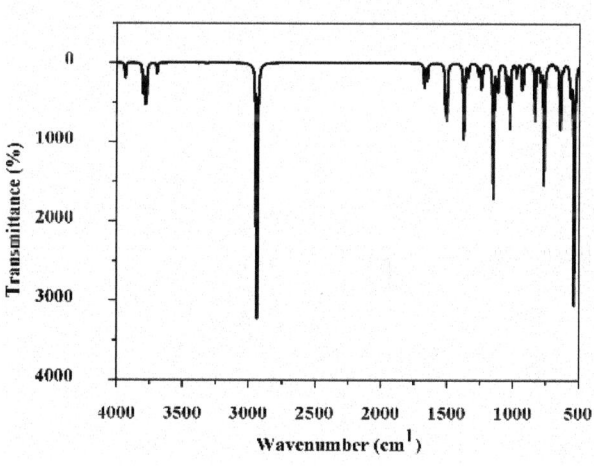

Fig. 3.4 Theoretically simulated FT-IR spectrum of LGHSQ

with other groups. Due to the carbonyl group interaction, inter and intramolecular hydrogen bonding are formed.

3.3.5 HYPERPOLARIZABILITY

The incident electromagnetic fields have been changed into higher order fields by means of frequency, phase and amplitude, from this then the electromagnetic fields transfer to various medium. The polarization of the molecule can be measured systematically by an induced dipole moment (μ). To evaluate the dipolar interaction by the whole external electric field, the Taylor series expansion should be working in the components of an electric field in case of poor polarization circumstance. The molecular first polarizability (β_0) and its associates (β, α_0 and $\Delta\alpha$) of LGHSQ single crystal has been calculated. In this calculation, the Lee-Yang-Parr correlation functional and Becke three parameter hybrid exchange functional were operated with help of B3LYP/6-311++G (d, p) basis set respectively.

The molecular dipole moment, the isotropic polarizability, the directionally dependent polarizability using the x, y, z components are defined as

$$\mu = (\mu_x + \mu_y + \mu_z)^{1/2}$$

$$\alpha_0 = (\alpha_{xx} + \alpha_{yy} + \alpha_{zz})/3$$

$$\Delta_\alpha = 2^{-1/2}[(\alpha_{xx}-\alpha_{yy})^2 + (\alpha_{yy} - \alpha_{zz})^2 + (\alpha_{zz} - \alpha_{xx})^2 + 6\alpha^2 xz]^{1/2}$$

The gain from Gaussian '09 contributes 10 components of the molecular first polarizability and is calculated using the following equation:

$$\beta_i = \beta_{iii} + \frac{1}{3}\Sigma(\beta_{ijj} + \beta_{iji} + \beta_{jji}) \; (i \neq j)$$

The total value of β can be calculated by x, y and z elemental molecular first polarizability.

$$\beta_{tot} = [\beta^2_x + \beta^2_y + \beta^2_z]^{1/2}$$

Table 3.4: The electric dipole moment μ, the average polarizability $α_{tot}$ and First order hyperpolarizability $β_{tot}$ for LGHSQ

Dipole Moment in Debye	
$μ_x$	-3.1229
$μ_y$	6.02302
$μ_z$	3.06739
$μ_{tot}$	2.4428
Polarizability in esu	
$α_{xx}$	46.1483
$α_{xy}$	8.02020
$α_{yy}$	18.3609
$α_{xz}$	6.35848
$α_{yz}$	4.52580
$α_{zz}$	12.2536
$α_o$	68.593×10^{-24}
Hyperpolarizability in esu	
$β_{xxx}$	-77.7521
$β_{xxy}$	-26.1051
$β_{xyy}$	-8.32824
$β_{yyy}$	-2.00129
$β_{xxz}$	-2.98801
$β_{xyz}$	-0.248184
$β_{yyz}$	-6.23642
$β_{xzz}$	7.09967
$β_{yzz}$	2.0869
$β_{zzz}$	-0.364416
$β_{tot}$	7.3144×10^{-30}

The following expression is used to determine the β value from Gaussian '09 output.

$$\beta = A = \pi r^2[(\beta_{xxx} + \beta_{xyy} + \beta_{xzz})^2 + (\beta_{yyy} + \beta_{yzz} + \beta_{yxx})^2 + (\beta_{zzz} + \beta_{zxx} + \beta_{zyy})^2]^{1/2}$$

The elements of the electric dipole moment (μ), Polarizability (α) and molecular first polarizability (β) values of LGHSQ have been calculated and presented in table 3.4. The determined first order hyperpolarizability of LGHSQ is 7.3144X 10^{-30} e.s.u. The static dipole moment (μ) is 2.4428 Debye, which has been compared with the reported values of urea (Ren'e Csuka et al 2012).

3.3.6 HOMO AND LUMO ANALYSIS

The investigation of the wave function specifies that the electron captivation corresponds to the transition from the ground state to the first excited state. The frontier molecular orbitals investigation is the promising techniques to found the molecular interaction between the molecules. The HOMO value can be act as an electron donor and the LUMO value can be act as electron acceptor respectively. From, the previous explanation the HOMO energy is directly connected to ionisation potential energy and the LUMO energy is directly connected to electron affinity. The eigenvalue of HOMO and LUMO and their energy gap reflect the chemical activity of the molecule. The calculated energies

E_{HOMO} is -5.0302 eV, E_{LUMO} is -0.5240 eV and ΔE is 4.5062 eV. is shown in Fig. 3.5. LGHSQ single crystal has an acceptable HOMO and LUMO level for NLO applications because of the relatively low energy gap values (Tagmatarchis N et. al. 2012).

HOMO energy; E_{HOMO} = - 5.0302 eV

LUMO energy; E_{LUMO} = - 0.5240 eV

HOMO–LUMO energy gap, $\Delta E_{Gap} = E_{HOMO} - E_{LUMO} = 4.5062$ eV

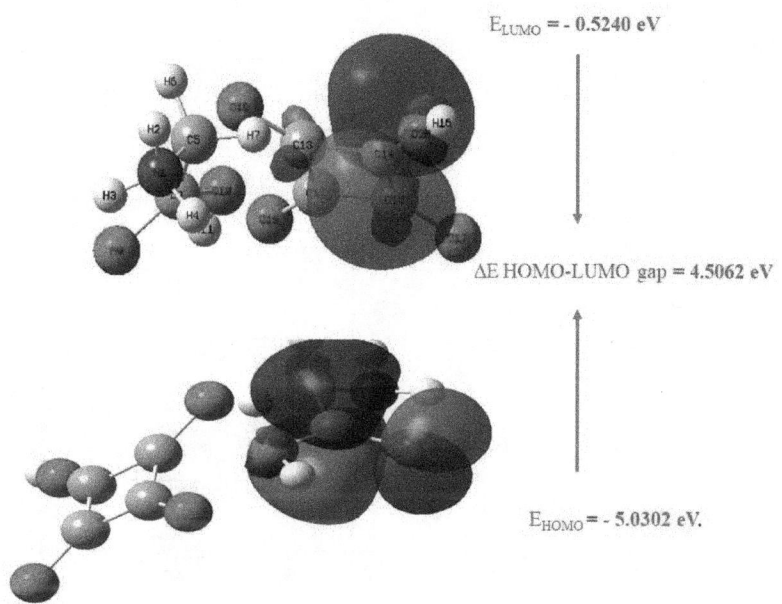

Fig. 3.5 HOMO – LUMO plot of LGHSQ at B3LYP/6-311++ G (d, p)

Table: 3.5: Calculated electronic and energies of LGHSQ using B3LYP/6-311++G (d, p) level

Electronic parameter	Calculated values
E_{HOMO} (eV)	-5.0302
E_{LUMO} (eV)	-0.5240
$\Delta E_{HOMO} - E_{LUMO}$ (eV)	4.5062
Ionization Potential (IP)	0.5240
Electron Affinity (EA)	5.0302
Chemical Hardness (η)	2.2531
Softness (s)	0.2219
Electronegativity (χ)	2.7771
Chemical Potential (μ)	-2.7771
Electrophilicity index (ω)	1.7114

The approximated HOMO-LUMO energy, hardness, softness, electronegativity, chemical potential and electrophilicity index of LGHSQ are approximate by DFT (B3LYP/6-311 ++ G (d, p)) basis with their respective values are listed in table 3.5.

3.3.7 THERMODYNAMICAL PROPERTIES

From, the thermodynamic functional analysis can be used to calculate the values of thermodynamical properties, thermodynamic quantities in particular heat capacity, entropy and enthalpy changes for synthesized LGHSQ single crystal have been measured by the vibrational analysis at B3LYP level with the 6–311++G (d, p) basis set in the range from 100K to 1000K and it is shown in Table.3.4. From this Table we concluded that those thermodynamic functions directly proportional to the temperature. Then, the optimized atomic vibrational intensities occur with temperature, the thermodynamic functions are optimized with temperature. The thermodynamic equations between the enthalpy, entropy and heat capacity, the temperatures will be fitting with the aid of quadratic formulas. The corresponding fitting factors (R^2) for LGHSQ single crystal are observed, and the vales are 0.9969, 0.9977 and 0.9999. The correlation graphs are shown in Fig. (3.6, 3.7 and 3.8). The corresponding fitting equations, can be explained as follow,

$$Cp^0{}_m = 0.442x^2 + 68.532T + 24.08 \times 10^{-5} T^2 \ (R^2 = 0.9969)$$

$$S^0{}_m = 0.337x^2 + 292.85T + 34.35 \times 10^{-5} T^2 \ (R^2 = 0.9977)$$

$$\Delta H^0{}_m = 0.447x^2 + -0.047T + 3.1791x \ 10^{-5} \ T^2 (R^2 = 0.9999)$$

The thermodynamic data for LGHSQ single crystal are useful for advanced study. The Gibbs functions are connected to the thermodynamic energies; they may be determined from thermodynamic data and the order of chemical reactions was also determined in line with the second law of thermodynamics from the same. The gaseous state and the liquid state have been used to conduct the whole thermodynamic computations.

3.3.8 OPTICAL STUDIES

The determination of exact optical nature of the single crystal helps to identify its extended credibility for optoelectronics device and non-linear optical (NLO) applications (M. Saravanan et al 2014). The optically transparent single crystal is attenuated by several external, and internal crystal orientation of the system (V. Pahurkar et al 2016, M.J. Weber et al 2002) and the physical parameters such as grain boundaries, vacancy, voids, inclusions, impurities, striations (R. Shaikh et al 2018). Optical absorption spectrum of the LGHSQ single crystal is shown in Fig.3.9. The optical band gab energy value is found to be 4.62 eV is shown in Fig.3.10. The spectrum reveals that the transmittance of 1 mm thickness LGHSQ crystal is ~ 90% in entire range of spectrum, further it is observed that the transmission of crystal sharply falls to minimum value in lower wavelength region. The sharp fall in transmittance identifies the cut-off wavelength of LGHSQ crystal at 286 nm (facilitated by n to π transition) and confirms the homogenous optical density of crystal.

The L-Glycinium Hydrogen Squarate (LGHSQ) single crystal with high optical transmittance and lower cut-off wavelength is the essential devices for transmission of second and third harmonic optical signals and UV-tunable lasers etc., (A. Shanthi et al 2013, M. Anis et al 2018). Thus, the influence of Extinction coefficient (k), refractive index (n), and reflectance (R) of LGHSQ crystal has been examined using the measured transmittance data was used shown in Figs 3.11 to 3.14. The material with low refractive index is directly used in holographic data storage utilities also, they find huge application as antireflection coating material for solar device (R. Santhakumari et al 2011). Extinction coefficient give the reference of photon energy.

3.3.9 MULLIKEN POPULATION ANALYSIS:

The natural population analysis (NPA) of the L-glycinium hydrogen squarate (LGHSQ) molecule was found by Mulliken Population Analysis (MPA) using B3LYP 6–311++G (d, p) basis set. Mulliken atomic charge calculation supports to understand the chemical potential and ionization potential of the molecule. The atomic charges are affecting the polarizability, electronic structure, dipole moment and other molecular properties of the system. Charge distribution of glycinium hydrogen squarate shows all the hydrogen atoms are positively charged, while the magnitudes of the atomic charges on carbon atoms for the glycinium hydrogen squarate (LGHSQ) compound as seen in Fig.3.15. were observed to be both positive and negative ranging from −0.6572 to 0.4768. When we compare the atomic charge for the all atoms, we note that the maximum atomic charge is obtained for H3. This is owing to negatively charged carbon C1 atom attached. Moreover, the oxygen atom exhibits the large negative charge. The presence of negative charge on the oxygen atom and net positive charge on the

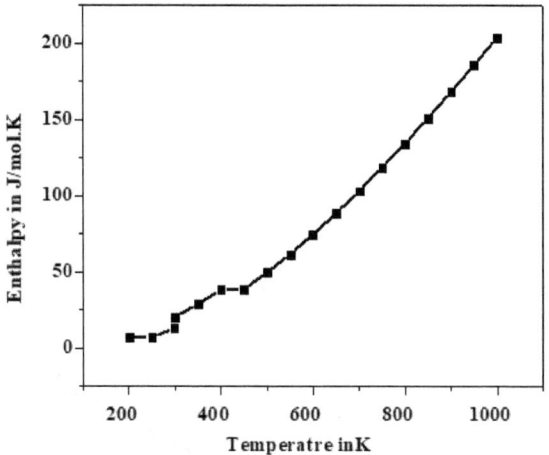

Fig.3.6 Variation of Enthalpy with temperature

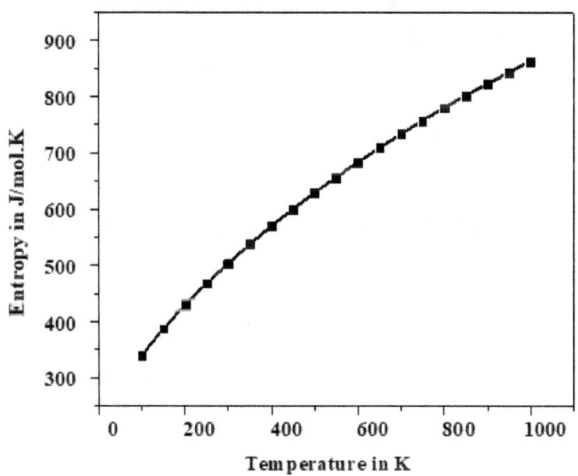

Fig. 3.7 Variation of Entropy with temperature

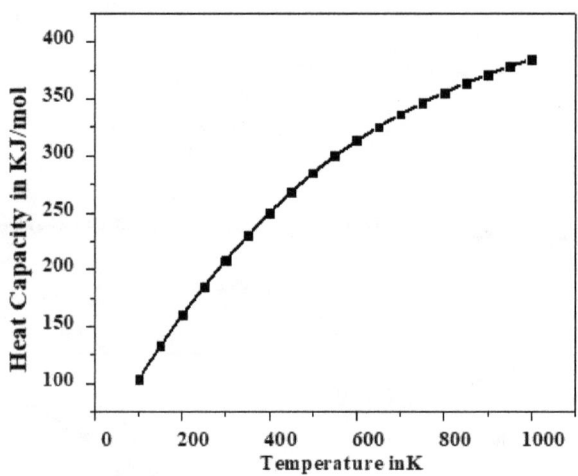

Fig. 3.8 Variation of heat capacity with temperature

Table 3.6: Thermodynamic properties at different temperatures by B3LYP level for LGHSQ

T (K)	S (J/mol.K)	Cp (J/mol.K)	ΔH (kJ/mol)
100	339.953	103.951	6.953
200	429.831	160.012	20.241
300	504.057	208.314	38.711
400	569.863	250.04	61.685
500	629.526	284.867	88.487
600	684.07	313.359	118.447
700	734.182	336.649	150.986
800	780.429	355.841	185.64
900	823.291	371.81	222.047
1000	863.179	385.211	259.917

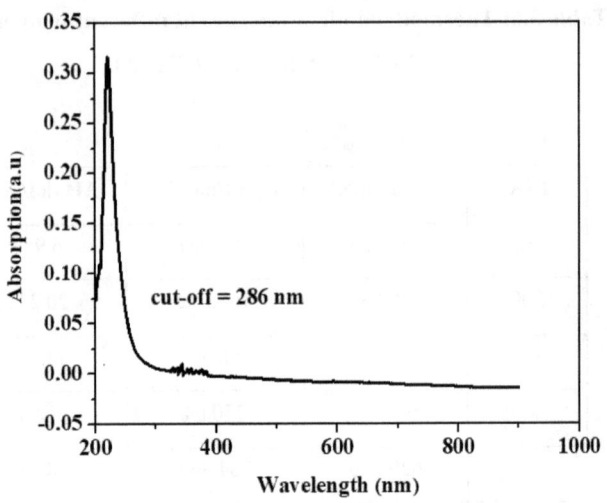

Fig. 3.9 UV- Vis absorption spectrum of LGHSQ crystal

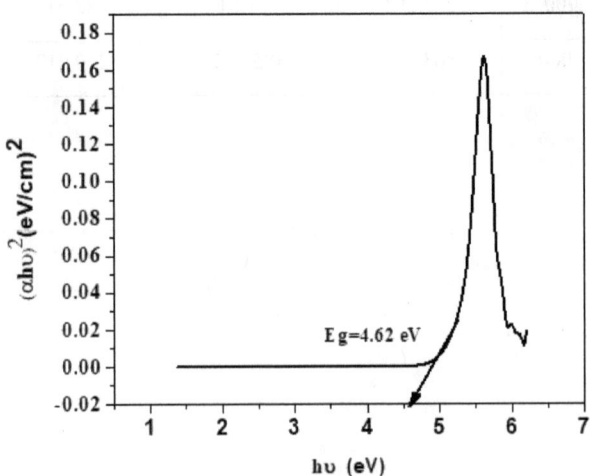

Fig. 3.10 UV- Vis bandgap plot of LGHSQ crystal

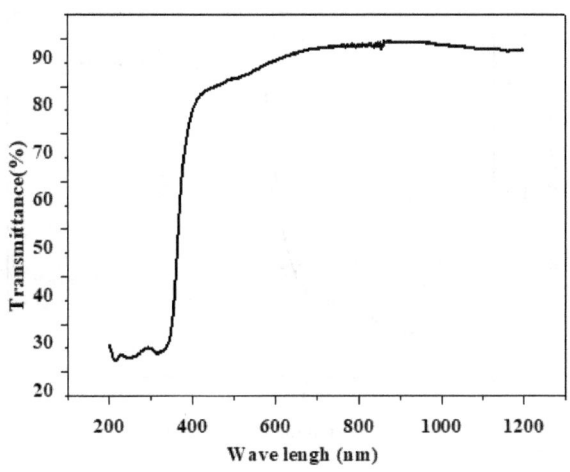

Fig. 3. 11 UV-Vis transmission spectrum of LGHSQ crystal

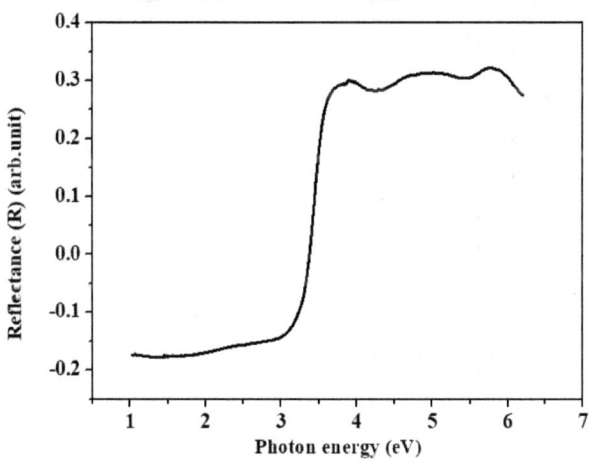

Fig. 3. 12 UV- Vis reflective index of LGHSQ crystal

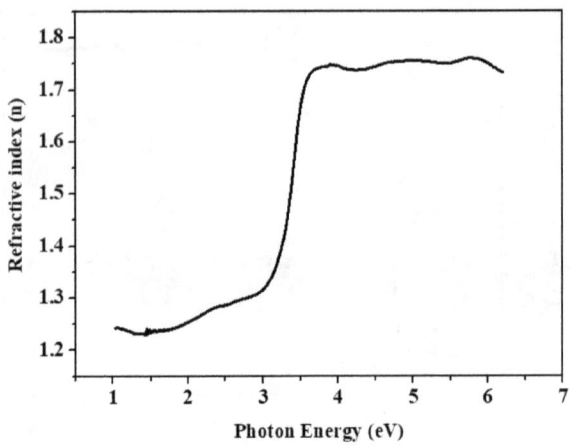

Fig. 3. 13 UV- Vis refractive index of LGHSQ crystal

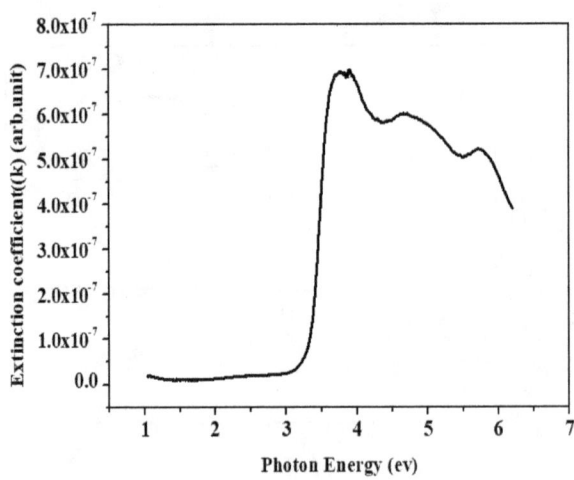

Fig. 3. 14 UV-Vis extinction coefficient of LGHSQ crystal

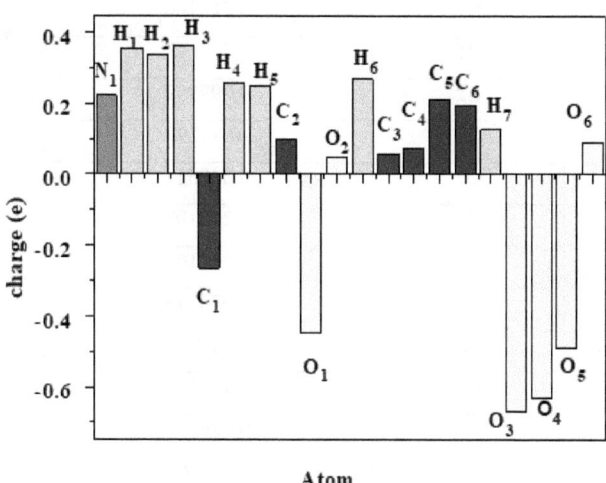

Fig. 3.15 Mulliken atomic charges of LGHSQ single crystal

Table 3.7 Mulliken atomic charges of LGHSQ single crystal

Atoms	Mulliken atomic charge
N1	0.221761
H1	0.353429
H2	0.337748
H3	0.362496
C1	-0.26596
H4	0.256166
H5	0.24897
C2	0.097999
O1	-0.44799
O2	0.048511
H6	0.268197
C3	0.057168
C4	0.073714
C5	0.210508
C6	0.194353
H7	0.126253
O3	-0.66763
O4	-0.63138
O5	-0.48915
O6	0.088368
N3	-0.8601
H13	0.4192
H14	0.4081
N4	-0.8601
H15	0.4032
H16	0.4081
C7	0.4624
C8	0.2903
C9	0.3064

This consequence proposes that the oxygen atoms act as lone pair donor and the charge transfer takes place from oxygen to carbon due to electron accepting substitutions at that position in our molecule. Mulliken atomic charges of LGHSQ single crystal are listed in Table 3.7.

3.3.10 NATURAL BONDING ORBITAL (NBO) ANALYSI

The NBO investigation is a useful method for deciphering the interactions between unoccupied acceptor atoms and occupied donor atoms in molecule. Table 3.8 shows the important relationship between acceptors and donors. The DFT/B3LYP6-311++ G (d, p) set was used to complete the NBO analysis of the complex. The intense intermolecular hyper conjugative interaction of π-electrons among the aromatic ring structure stabilises the LGHSQ compound. The orbital overlap between (C27-C28) and (C27-C28) determines the intermolecular interaction. The aromatic ring of the LGHSQ compound is balanced out as a result of these intermolecular charge transfers, resulting in a stabilization energy of 10.55 Kcal/mol at the B3LYP6-311++ G (d, p) stage.

Intermolecular interactions between antibonding with LP and stabilisation energy of 33.92 Kcal/mol were found in π* (N18-C20) and n1 (N21) indicating large delocalisation. The maximum hyper conjugative interactions occurs between the C27-C32 antibonding molecule, resulting in an 313.57 Kcal/mol stabilisation capacity.

3.3.11 MICROHARDNESS TEST

Microhardness plays a key role in device fabrication. Vicker's hardness measurements were made on the prominent plane using Leitz–Wetzler hardness tester fitted with a diamond pyramidal indenter. The LGHSQ crystal with a clear surface has been tested for indentation at room temperature by Vickers hardness test. Loads of different weights such as 25, 50, 100 g have been applied to all loads applied and the indentation time has been kept as 10s. The Vicker's microhardness number was determined using the expression (Karan et al 2005),

$$Hv = 1.8544 \ (P/d^2) \ Kg \ mm^{-2}$$

Where P and d show the load applied in kg and the average diagonal length of the indentation in mm. The plot between the load applied P and the corresponding hardness number HV is shown in Fig. 3.16. The good hardness number elevation in relation to the load describes the material high mechanical strength. The hardness of the material is guaranteed from the plot.

Table 3.8: Second order perturbation theory analysis of Fock Matrix in NBO for LGHSQ with 6-311++G (d, p) basis set

Donor (i)	Type	Acceptor (i)	Type	$E(2)^a$ (k cal/mol)	$E(j)-E(i)^b$ (a.u)	$F(I,j)^c$ (a.u)
C3-C4	σ	N11-O13	σ*	18.03	1.12	0.127
C4-C5	π	C3-O14	π*	26.72	0.25	0.079
N6-H8	π	C4-C5	π*	10.48	1.29	0.104
N11-O12	π	C4-C5	π*	27.49	0.33	0.091
N1	nl	C2-O7	π*	65.01	0.27	0.119
N1	nl	C3-O14	π*	89.22	0.25	0.134
N6	nl	C2-O7	π*	61.73	0.24	0.111
N6	nl	C4-C5	π*	43.06	0.25	0.097
O7	nl	C2	σ*	19.19	1.56	0.155
O7	π	N1-C2	σ*	28.99	0.63	0.123
O7	π	C2-N6	σ*	28.62	0.64	0.123
O12	π	N11-O12	π*	12.48	0.08	0.040
O13	π	N11-O12	π*	15.36	0.30	0.082
O13	π	O13-H23	σ*	51.34	1.28	0.235
O14	nl	C3	σ*	13.00	1.59	0.129
O14	π	N1-C3	σ*	20.22	0.83	0.127
O14	π	N11-O12	π*	172.46	0.09	0.138
C3-O14	π*	C4-C5	π*	193.35	0.03	0.095
N11-O12	π*	C4-C5	π*	10.90	0.19	0.045
O13-H23	σ*	H23	σ*	82.54	0.05	0.136

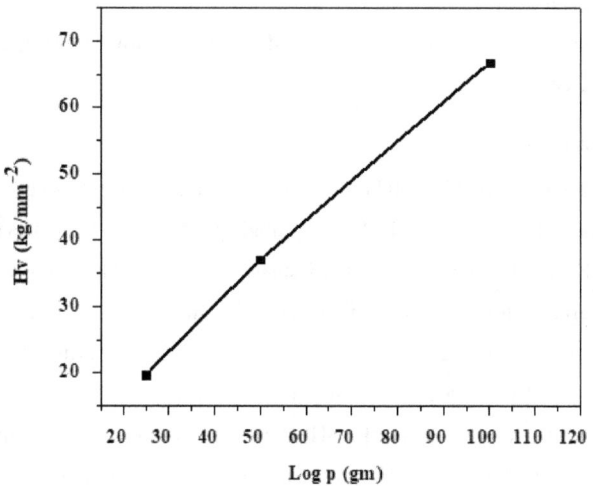

Fig. 3.16 Vickers hardness number Vs applied load of LGHSQ

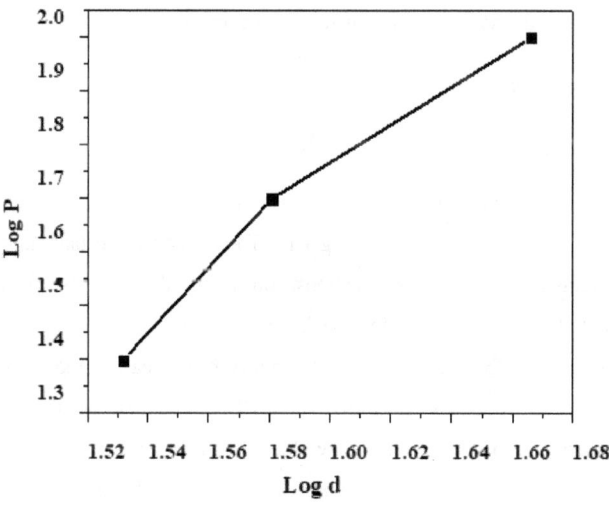

Fig. 3.17 log p Vs log d plot of LGHSQ

between log P and log d in Fig. 3.18. The value of Meyer's index 'n' can be calculated from the slope of the graph between log P and log d, shown in the Fig. 3.17. The slope value (n) was found to be 1.6. According to Onitsch, we could say that the grown LGHSQ crystal comes under soft material category

3.3.12 Thermal analysis

The thermogravimetric (TG) and differential thermal analysis (DTA) has been used to investigated the thermal behaviour of the LGHSQ single crystal. Fig.3.18. shows the TG-DTA spectrum of LGHSQ crystal. Thermogravimetric analysis (TGA) and Differential thermogram analysis (DTA) gives information regarding the phase transition and different stages of decomposition of the grown single crystal. The thermal analysis has been used to detect the weight loss (TGA) and melting point of the material, decomposition point (DTA) of the LGHSQ single crystal. The DTA curve of LGHSQ crystal has a major endothermic at 108° C. The endothermic transition of the LGHSQ crystal's DTA curve occurs between 235° C and 266° C. This explanation claims that the LAHSQ material remains stable up to at 206°C. The melting point of the LGHSQ single crystal is represented by the sharp endothermic value at 235° C. Another important observation also been observed in this investigation, there is no phase transition and colour change till the material melts. It helps to enhances the temperature range of the compound and the utility grown single crystal has been used for the NLO application.

3.3.13 Dielectric studies

The dielectric constant graph shows the variation of dielectric constant for the *log* frequency range of 1.5-6.5 Hz is shown in Fig.3.19. The dielectric constant steeply decreased with increasing frequency until 300 Hz and almost constant for the higher frequency sides. The dielectric constant at lower frequency follows linear relation with space-charge polarization. The polarisation value will be decreases then, it exhibits the decrease in the value of dielectric constant with increasing frequency (L. Guilbert et al 1998). The dielectric loss studies explain the ability of materials, it helps to convert the electromagnetic energy into heat energy. The variation of dielectric loss (*tan δ*) with frequency can be shown in Fig.3.20. The dielectric loss decreases with increase of frequency. It indicates low dissipation of energy in the form of heat (M. Vimalan et al 2010), which further confirms the quality of LGHSQ single crystal. The low value of dielectric constant and dielectric loss at the higher frequency suggests LGHSQ single crystal possess the less defects and it was suitable for NLO device applications. From Fig.3.21.,

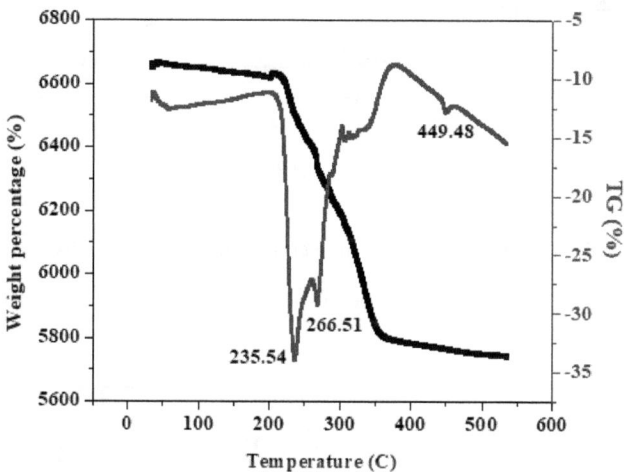

Fig. 3.18 TG-DTA plot of LGHSQ crystal

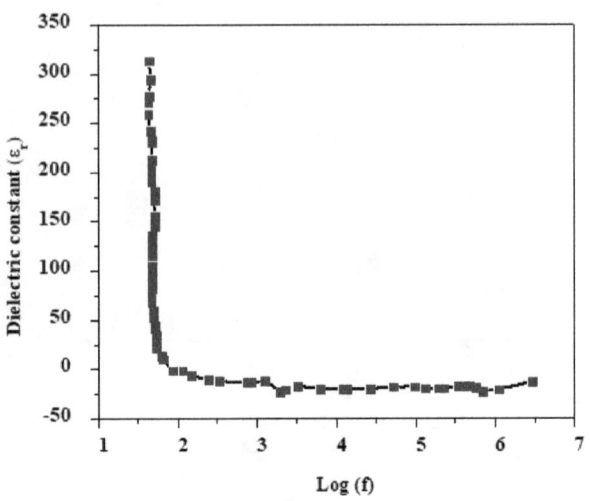

Fig. 3.19 Variation of dielectric constant of LGHSQ

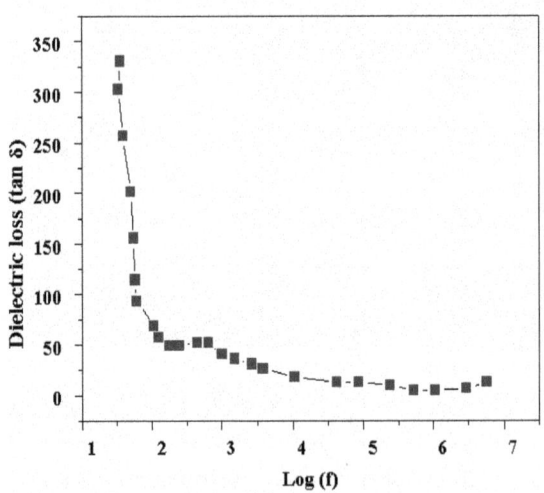

Fig. 3.20 Variation of dielectric loss of LGHSQ

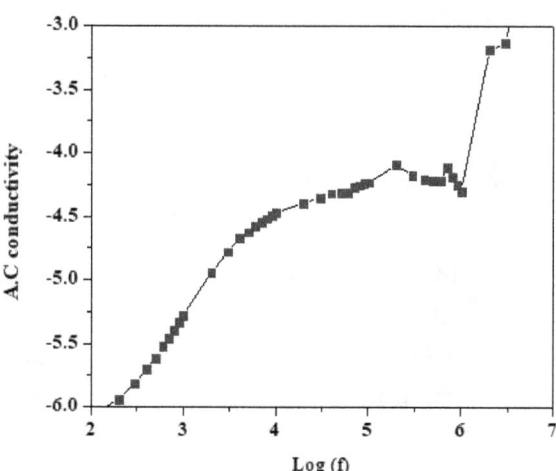

Fig. 3.21 Frequency dependence of AC Conductivity

it can be clearly seen that the value of σ_{ac} is very low up to 6 MHz and increase with increase of frequency and can be explained on the basis of frequency power law.

3.3.14 PHOTOLUMINESCENCE ANALYSIS

The photoluminescence (PL) studies depend on the excitation energy as well as the intensity of the incident beam. The LGHSQ single crystal has been photo-excited with an excitation wavelength of 450 nm and the EM spectrum was recorded with no delay in the range of 200 to 800 nm. The EM (emission) spectrum shown in Fig.3.22. and it reveals that LGHSQ crystal has strong blue emission with the measured peaks centred at 487 nm. When electrons go from the excited state to the ground state there is a loss of vibrational energy. As a result, the emission spectrum is shifted to longer wavelength than the excitation spectrum. So, the PL spectrum of LGHSQ was the longer excitation wavelength. The unique nature of Photoluminescence peaks indicates that the LGHSQ single crystal has good optical quality. This stringent requirement for material, can be considered as NLO property will be in active (Mohd Anis et al 2014).

3.3.15 MOLECULAR ELECTROSTATIC POTENTIAL

The total electrostatic effect or field produced at a point in space around a molecule is due to the net charge distribution of electron and nuclei is termed as molecular electrostatic potential (MEP) V(r). It can be expressed as the following equation (P. Politzer and J.S. Murray et al 2002),

$$V(r) = \Sigma A \frac{ZA}{|RA-r|} - \int \frac{\rho(r) dr^F}{r'-r}$$

From the above equation where ZA is the charge on nucleus located at a distance of RA and $\rho(r)$ is the electron density. The Molecular Electrostatic Potential (MEP) is interrelated with the dipole moments, chemical reactivity and electronegativity of the molecule. It gives the visual inspection of the polarity of the molecule. The electrostatic potential region was taken between $-0.191 \times e^0$ and $0.191 \times e^0$. The MEP map is shown in Fig. 3.23. The electrostatic potential surface of the molecule can be characterized by different colour grading. The red value can be representing the most negative electrostatic potential region in a molecule, whereas the blue value can be representing the most positive electrostatic potential region and the region of zero potential represent green colour. The attraction of proton by the concentrated electron in the molecules can be obtained by the negative electrostatic potential energy. This result gives the information about the electrophilic attack or reactive sites of the molecular system.

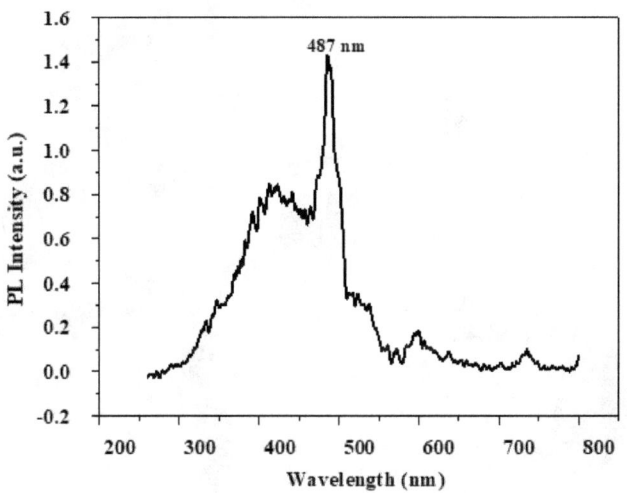

Fig. 3.22 Fluorescence emission spectrum of LGHSQ crystal

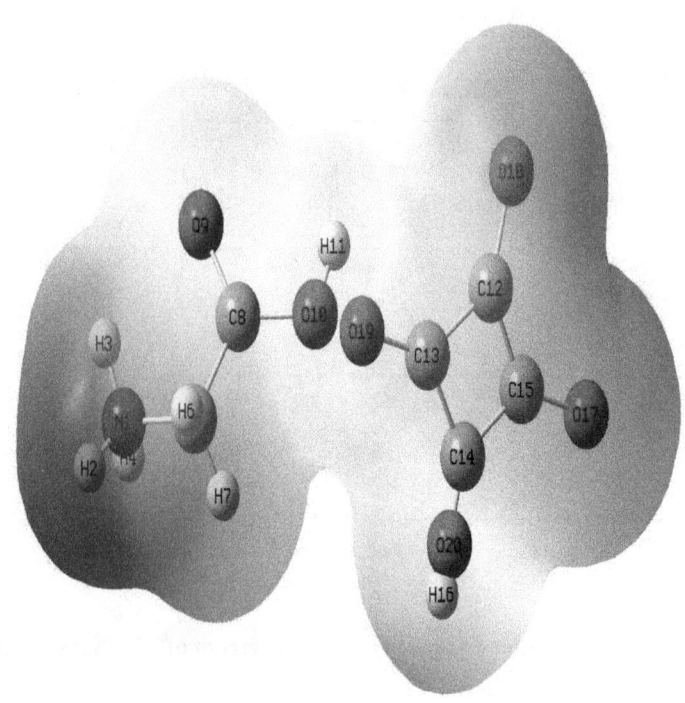

Fig. 3.23 Molecular Electrostatic Potential (MEP) of LGHSQ crystal

The repulsion of proton by atomic nuclei in the regions of low electron density excites, which can be obtained by positive electrostatic potential energy. This result gives the information about the nucleophilic attack or reactive sites of the molecular system. From this investigation helps to give the information about the region from where the compound can have intermolecular interaction in the system.

3.3.16 Second harmonic generation (SHG)

The frequency doubling (nonlinear) property of LGHSQ single crystal has been investigated by using the Kurtz-Perry powder technique (I. Hubert Joe et al 2009). This technique can be used to measure the Second Harmonic Generation (SHG) effective non-linearity of new materials relative to potassium Dihydrogen Phosphate (KDP). The LGHSQ single crystal is 1.55 times greater in second order efficiency (31 mV) when compared with the standard Potassium dihydrogen phosphate (KDP) material (20 mV). From this analysis to conclude that the LGHSQ crystal would be an optional material in SHG device fabrications.

3.4 CONCLUSION

The theoretical calculations for the LGHSQ single crystal were performed using the DFT method. A theoretical geometric structure for the molecule's bond lengths and bond angles is optimized, as is an assessment of the structure against similar compounds. The FT-IR spectroscopic examination of the LGHSQ molecule, experimentally and theoretically determined vibrational wavenumbers, are seen. The theoretical results showed overall agreement with the experimental record, but they also exhibited significant discrepancies. The molecular energy gap is uncovered via the research of border molecular orbitals. The estimated energy gap E =4.62 eV of the LGHSQ single crystal implies that it might be a viable NLO material. The value of the slope (n) was discovered to be 1.6 from Meyer's index. According to TG/DTA tests, the material is stable up to 206 degrees Celsius. The endothermic transition of the LGHSQ crystal's DTA curve occurs between the values of 235° C and 266° C. In addition to the compound's ionic conductivity, the electrical conductivity was also studied. The NBO and mulliken charge estimations were discussed. In terms of NLO efficiency, the conventional KDP material received a benefit of approximately 1.55 times the NLO efficiency of the control.

CHAPTER IV

EXPERIMENTAL AND COMPUTATIONAL STUDIES ON L- ARGININIUM 5-NITROURACILATE (LA5N) SINGLE CRYSTAL

4.1 INTRODUCTION

Motivated by numerous potential applications, an important effort has been devoted over the past few decades to the search for new materials for nonlinear optics. The varied performance of organic derivatives and adequate tailoring of these gave rise to new types of compounds that display higher nonlinear-optical susceptibilities. In contrast, organics exhibit significantly higher nonlinear coefficients, generally at the price of a reduced transparency domain and possible stability problems. The molecular structure of LA5N is represented in Fig. 2.1. It may be considered as a cyclic urea derivative, with a strong electron-accepting group (nitro). Owing to the presence of N-H bonds and of C=O groups, hydrogen bonding interactions are expected with protic solvents as well as with other LA5N molecules in the solid state. This potential is experimentally confirmed by both solubility and crystallographic packing interatomic-distance analysis. In principle the cyclic structure of LA5N is considered to be nonaromatic. Some interaction, however, is expected to occur between the π electrons of the C=C double bonds and the nonbonding electrons of the out-of-plane p_z orbital of the SP^2 hybridized nitrogen atoms belonging to the N-H groups. The nitro group may also interact with the electrons of the uracil ring. There is only one form of LA5N that is non centrosymmetric, has $P2_12_12_1$ space group (R.S. Gopalan et al 2004) and displays high nonlinear optical properties (G. Pucceti et al 1993). In this regard, the idea of realizing the non-linear optical potential of 5-nitrouracil in its new compounds is particularly attractive. The authors of (S.R. Domingos et al 2012) obtained the salt L-histidinium 5-nitrouracilate; and the presence of amino acid L-histidine chiral molecule inevitably leads to the formation of a non-centrosymmetric structure. So, we tried to obtain a L-Argininium 5-Nitrouracilate single crystal. So, it crystallizes in the polar space group $P2_1$ and displays highest non-linear optical properties among the L- Argininium salts. According to (M. Fleck et al 2014 and H.A. Petrosyan et al 2005), L- Argininium (as well as other amino acids, having a basic side chain) can form salts not only with singly, but also with a doubly charged cation. In the present work, we show the results of the structural investigation of the new monoclinic form of single crystal. In addition, the influence of the crystal structure on the non-linear optical properties has been studied.

4.2 SYNTHESIS AND GROWTH

Single crystals of LA5N were grown from supersaturated solution by slow evaporation technique. The substance of L- Arginine -$C_6H_{14}N_4O_2$ (Sigma–Aldrich, 99%) and 5-Nitrouracil - $C_4H_3N_3O_4$ (Sigma–Aldrich, 99%) were taken in equimolar ratio (1:1). The solution was mixed together to dissolve in water until we get a saturated solution. Then, the highly dissolved salt solution is filtered by whattman filter paper and poured in a beaker. The beaker is covered with a silver foil paper with a few small holes for evaporation of solvents. The apparatus is placed undisturbed till the seeds of sufficient size of the given L- Arginine and 5-Nitrouracil are obtained in the beaker. The LA5N single crystal was harvested within 35-40 days from the mother solution. The photograph of LA5N single crystal is shown in Fig.4.1.

4.3 RESULTS AND DISCUSSION

4.3.1 SINGLE CRYSTAL X-RAY DIFFRACTION

From the single crystal XRD data, it is found that the title compound belongs to the crystal system monoclinic with $P2_1$ space group and has the following cell dimensions; a =8.863(9) Å; b =9.965(6) Å; c =15.821(15) Å.

4.3.2 COMPUTATIONAL DETAILS

Quantum chemical density functional computations were carried out at the Becke3-Lee-Yang-parr (B3LYP) (A.D. Becke et al 1993) level with 6-311++G (d, p) basis set using Gaussian 09 W program package to get a clear knowledge of optimized parameters. The optimized molecular structure is used for the computation of vibrational frequencies, like IR intensities with the Gaussian 09 W software. The energy distribution from HOMO to LUMO was calculated using the Gauss 09 software program which served as a precursor to theoretically attain the band gap of the title molecule. These results were compared to the band gap obtained by theoretical calculation with respect to the UV-Vis spectrum.

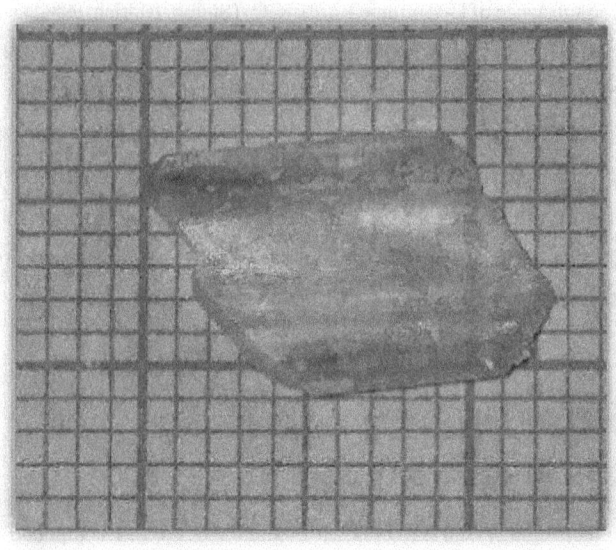

Fig. 4.1 Photograph of as grown LA5N single crystal

Table 4.1: Crystal parameters of LA5N single crystal

Empirical Formula	$C_{10}H_{19}N_7O_4$
Formula weight	288.27 g/mol
Crystal system	Monoclinic
Space group	$P2_1$
Unit cell dimensions	a = 8.863 (9)Å, $\alpha = 90°$ b = 9.965(6)Å, $\beta = 9.16°(15)$ c = 15.821(15)Å, $\gamma = 90°$
Cell volume	502.24 Å^3
Absorption coefficient	0.198 mm^{-1}
F (000)	695
Crystal size	0.630 x 0.13 x 0.20 mm^3
Theta range for data collection	4.254 to 25.145°
Limiting indices	-8<=h<=8, -16<=k<=16, -18<=l<=18
Reflections collected/unique	2475/ 2251
Completeness to theta = 23.92	96.5%
Refinement method	Full-matrix least-squares on F^2
Data/restraints/parameters	2745 / 1 / 125
Goodness-of-fit on F^2	1.524
Final R indices [I>2sigma(I)] 80	R1 = 0.0152, wR2 = 0.0547
R indices (all data)	R1 = 0.0748, wR2 = 0.0574

4.3.3 MOLECULAR GEOMETRY

The optimized molecular structure of the LA5N compound with the numbering scheme of the atoms is shown in Fig. 4.2. The bond parameters such as bond lengths and bond angles obtained from Gaussian 09 W. These theoretical results were compared with the experimental data (Bin Jiang et al 2009) of the title compound and are listed in Table 1. The title compound was found to have eight C-N bonds, nine C-H bonds, six C-C bonds, two C-O bonds, ten N-H bonds. The bond lengths N11-C18 (1.269 Å) and C5-N4 (1.380 Å) were found to be less than the other N6-C1 (1.235 Å) bonds, similarly C1-O8 (1.207 Å) and C18-C19 (1.652 Å) bond lengths were found to be lesser than the corresponding N6-H12 (.860Å) and N4-H13 (0.962 Å) bonds.

This result may be due to the fact that bond lengths in a double bond are smaller when compared to single bonds, as the double bonds are stronger in nature. The theoretical calculations were carried out as an isolated molecule in the gas phase, which leads to the deviation of the values calculated theoretically from experimentally observed data which was obtained from a single crystal.

The molecular structure along with numbering of atoms of is shown in the Fig. 4.2. Table 4.2 and 4.3 depicts the optimal structural parameters, which were calculated from the parameter study.

4.3.4 VIBRATIONAL ASSIGNMENTS

On the recorded FT-IR spectra, the vibrational spectral assignments were performed based on the theoretically predicted wavenumbers by density functional theory (DFT)of B3LYP/6- 311++G (d, p) method. Normally in all the heterocyclic compounds, the N-H stretching vibration occurs in the region of 3000-3500 cm^{-1}. The LA5N compound has only one NH_2 group and hence one symmetric and one asymmetric N-H stretching vibrations in NH_2 group are expected. The asymmetric stretching for the N-H has a magnitude higher than the symmetric stretching that appear from 3420-3500 cm^{-1}. The O-H stretching vibrations are in the region of 3500-3800 cm^{-1} (H. Endredi et al 2003). The bands between 1400 and 1700 cm^{-1} in the aromatic and hetero aromatic compounds are assigned to carbon vibrations. Frequency nearer to 1500 cm^{-1} indicates C=N bonds while frequency nearer 1200 cm^{-1} indicates the presence of C-N bonds (I. Hubert Joe et al 2009). The values obtained show fairly good agreement with the experimental spectra and it has been shown in Fig.4.4.

VIBRATIONS OF C-C GROUP

The ring stretching vibrations are very much important in the spectrum of benzene, carbon–carbon ring stretching vibrations occur in the region 1405-1600 cm^{-1}. In the present study, the C–C stretching vibrations for title molecule are found between 1600–1485 cm^{-1} in.

Fig. 4.2 Atomic numbering system adapted for ab initio computations of LA5N molecule

Table 4.2: Selected bond lengths of LA5N molecule

Parameters	Bond length in (Å)	
	Observed Bond Length	Calculated Bond Length
C1-C2	1.475	1.3965
C1-N6	1.387	1.3254
C5-O8	1.2070	1.9935
C1-O7	1.2150	1.2845
C2-N11	1.436	1.3987
C2-C3	1.350	1.3692
N16-C18	1.269	1.2457
N11-O18	1.220	1.9875
C5-N4	1.380	1.3952
C21-N23	1.470	1.4652
N15-C17	1.468	1.4580
N4-H13	0.860	0.7985
C21-H34	1.070	1.0632
N24-H28	1.000	0.9952
N6-H12	0.860	0.7925
C3-N4	1.420	1.396
N4-H13	0.962	0.965
N6-C1	1.235	1.112
N11-O10	0.996	0.992
N11-O9	0.996	0.992
N15-H25	0.965	0.999
C17-N16	1.235	1.114
C17-N24	1.472	1.396
C18-C19	1.652	1.452
C19-C20	1.742	1.698
C20-C21	2.542	2.226

Table 4.3: Selected bond angles of LA5N molecule

Parameters	Bond Angle in (°)	
	Observed Bond Angle	Calculated Bond Angle
C1-N6-H12	128.564	129.365
C1-C2-C3	135.928	136.325
C2-N11-O10	142.964	142.365
C2-C3-N4	107.726	107.265
N4-C5-O8	110.111	109.365
N11-C2-C3	95.730	98.412
H12-N6-C1	111.171	110.321
N6-C1-O7	122.508	121.254
N6-C1-C2	114.095	113.254
O7-C1-C2	133.140	132.254
C1-C2-N11	119.773	120
C2-C3-H14	130.120	129.324
C2-C3-N4	117.329	115.365
C3-N4-C5	125.718	13.541
C18-C19-C20	175.612	174.221
N15-C17-N16	128.546	128.235
C17-N24-H28	132.170	133.241
C18-C19-C20	156.606	157.247
C22-C21-N23	92.280	93.158
C20-C21-N23	173.234	174.254
N16-C8-C19	110.215	109.142
H27-N24-H28	134.588	134.111
C19-C20-H23	135.392	136.021
C17-N15-H26	116.280	115.201

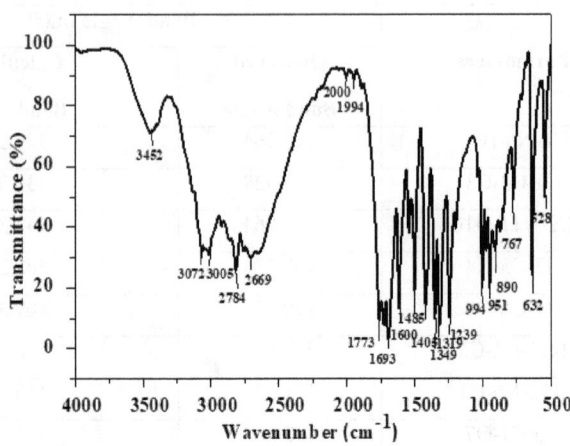

Fig. 4.3 Experimentally obtained FT-IR spectrum of LA5N

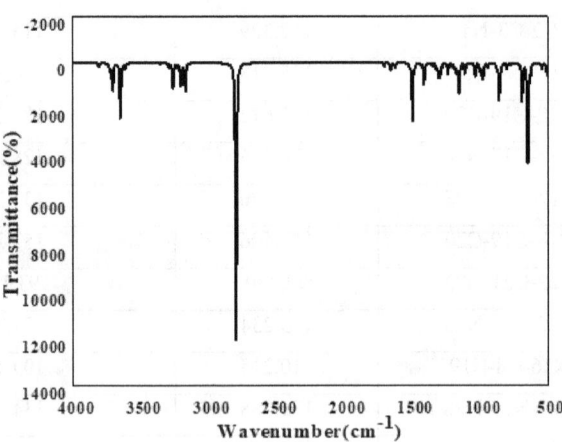

Fig. 4.4 Theoretically simulated FT-IR spectrum of LA5N

both FT-IR spectra. The computed wavenumbers at 1610, 1606, 1494, 1455, 1377 and 1339 cm^{-1} by B3LYP method assigned C–C stretching vibrations. The C–C–C in-plane bending bands always occur between the value 1000–600 cm-1 (A. Fu, et al 2003). The bands observed at 767 cm $^{-1}$ in FT-IR spectrum are assigned to C–C–C deformations. The theoretically computed value at 744 and 731cm^{-1} gives excellent agreement with experimental data.

VIBRATIONS OF C-N AND N-H GROUP

The N–H stretching vibrations appears as a strong broad band in the region 3500–3300 cm^{-1} (A. Altun et al 2003 and L. J. Bellamy et al 1975). In the present case, the band observed at 3072 cm–1 in FT-IR spectrum assigned as N–H stretching vibration. The C–N stretching vibration coupled with NH, is moderately to strongly active in the region 1349-1239 cm^{-1}. (A. Spire et al 2000) have observed a band at 1319 cm^{-1} in the IR spectrum as this C–N stretching mode. For the title compound, the C–N stretching modes are observed as a medium band at 1319 cm^{-1} in FT-IR spectrum. In N–H in-plane bending (CNH) vibration is observed at 1600 cm^{-1} in FT-IR and computed wavenumber at 1623 cm^{-1} in B3LYP method, this is good agreement with experimental findings. The N–H out-of plane bending vibration is observed at 1106 cm^{-1} in in calculated B3LYP/6-311(d, p).

VIBRATIONS OF C-O GROUP

The C-O wavenumbers will appear within the range of 1850-1600cm^{-1}. In this study, the experimental C-O vibration is observed at 1747 cm^{-1} in FT-IR which is comparable to the calculated B3LYP vibrational frequency method. The stretching C-O vibrations are found at 1773, 1659 cm^{-1} in FT-IR spectrum are assigned to C-O stretching vibrations are very good agreement with the literature values (D. H. Wiffen et al 1955). Moreover the calculated wavenumbers by B3LYP/6-311++ G (d, p) method are also well consistent with the experimental values. The in-plane and out-of-plane C-O deformations are expected in the regions 890-632 cm^{-1} and 632-58 cm^{-1} respectively.

VIBRATIONS OF N-O GROUP

In nitro compounds the most characteristic bands are due to NO$_2$ stretching vibrations, which are the two most important group wavenumbers, not only because of the spectral position but also for the strong intensity (N.P.G. Roeges, et al 1994). In nitro compounds the antisymmetric NO_2 stretching vibrations are located in the region 1485 cm^{-1}. The symmetric NO_2 stretching vibrations are expected in the region 1349 cm^{-1}. In the present case, the bands observed at 1600 and 1485 cm^{-1} in the FT-IR spectrum. Theoretically, the vibrations are observed in the 1519–1093 cm^{-1} region.

VIBRATIONS OF C-H GROUP

In general, the vibrational stretching mode of asymmetric and symmetric of C-H group occur about 2784 cm^{-1} and 2669 cm^{-1} (K. Mohana Priyadarshini et al 2014). The C-H asymmetric stretching vibration is perceived as a weak IR band at 2784 cm^{-1}. The Infrared band is appeared at 3005 cm^{-1} as a C-H symmetric stretching mode. The vibrations of CH_2 twist are appeared in the frequency range 1239-994 cm^{-1} and CH_2-rock vibrations existed in the region of frequencies 994-776 cm^{-1} (N.B. Clothup et al 1990). CH_2 wagging is perceived in IR at 1239 cm^{-1}. CH inplane bending and CH outplane bending appears in the vibrational frequency range 1349–1000 cm^{-1} and 951-632 cm^{-1} respectively (R.M. Silverstain et al 2005). In LA5N, CH in plane bending is appeared in FT-IR radiation at 994 cm^{-1}. Also, CH out plane bending is existed in IR at 767 cm^{-1}. C-H stretching, in plane bending and out of plane bending vibrational modes were possible in the title compound. Generally, the C-H stretching characteristic vibrational modes of heterocyclic aromatic compounds fall in the wave number 3072-3005cm^{-1}. In experimental C-H stretching vibration of benzene and pyrazine ring were observed as a strong peak at 3072 and 3000cm^{-1} in FTIR spectrum. In FTIR the observed ring C-H stretching vibrational frequency was shifted towards the higher wavenumber, which can be correlated to the presence of intermolecular interactions and hyper-conjugation interactions (S. Premkumar et al 2016). Usually, the aromatic C-H in plane bending vibrational modes contribute to the multiple peaks in the region 1600-1405 and 1239-994cm^{-1}, which sometimes interact with the ring stretching vibration (G. Socrates et al 2004).

4.3.5 HYPERPOLARIZABILITY

The first hyperpolarizability (β), the mean polarizability (α), the anisotropy of the polarizability (μ) and the total static dipole moment (l) of LA5N crystal have been predicted by using the DFT/B3LYP/6-311++G (d, p) method. First hyperpolarizability is a third rank tensor that can be described by 3x3x3 matrix. The 27 components of 3D matrix can be reduced to 10 components due to the Kleinmann symmetry (D.A. Kleinman et al 1962 and H. Tanak et al 2011). The output from Gaussian 09 provides 10 components of this matrix as β_{xxx}, β_{yxx}, β_{xyy}, β_{yyy}, β_{zxx}, β_{xyz}, β_{zyy}, β_{xzz}, β_{yzz}, β_{zzz}, respectively. The components of the first hyperpolarizability can be calculated using the following equations.

$$\beta_{total} = (\beta^2 x + \beta^2 y + \beta^2 z)$$

were,

$$\beta_x = \beta_{xxx} + \beta_{xyy} + \beta_{xzz}$$

$$\beta_y = \beta_{yyy} + \beta_{yzz} + \beta_{yxx}$$

$$\beta_z = \beta_{zzz} + \beta_{zxx} + \beta_{zyy}$$

The equations for calculating the magnitude of mean polarizability, anisotropy of the polarizability and total static dipole moment are defined as follows:

$$\alpha = 1/3\ (\alpha_{xx} + \alpha_{yy} + \alpha_{zz})$$

The elements of the electric dipole moment (μ), Polarizability (α) and molecular first polarizability (β) values of LA5N have been calculated and presented in Table 4.4. The determined first order hyperpolarizability of LA5N is 7.5735X 10^{-30} e.s.u. The static dipole moment (μ) is 2.2043 Debye, which has been compared with the reported values of urea (Ren'e Csuka et al 2012).

4.3.6 HOMO AND LUMO ANALYSIS

Frontier Molecular Orbitals (FMOs) play a vital role in the optical and electric properties, as well as in quantum chemistry and UV-Vis spectra and is used to explicate several types of reactions in conjugated systems (K. Fukui et al 1952). Moreover, eigenvalues of the LUMO, HOMO and their energy gap replicate the chemical reactivity of the molecule and the bioactivity from Intramolecular Charge Transfer (ICT) can be explained by the energy gap (L. Padmaja et al 2009 and C. Ravikumar et al 2008). Transition state transition of π-π* type is observed with regard to the molecular orbital theory due to the interaction between HOMO and LUMO orbital of a structure. The very strong band observed at 431 nm. The band gap energy was estimated using the formula, $E = hc/\lambda$. Here h and c are constant; λ is the cut off wavelength. The chemical reactivity and the kinetic stability of the molecule are described by the frontier orbital gap. A soft molecule with a small frontier orbital gap has a high chemical

Table 4.4: The electric dipole moment μ, the average polarizability α_{tot} and first order hyperpolarizability β_{tot} for LA5N molecule

Dipole Moment in Debye	
μ_x	3.4587
μ_y	2.4127
μ_z	-1.0124
μ_{tot}	2.2043
Polarizability in esu	
α_{xx}	36.25
α_{xy}	5.28
α_{yy}	28.13
α_{xz}	4.77
α_{yz}	2.15
α_{zz}	27.22
α_o	30.53333×10^{-24}
Hyperpolarizability in esu	
β_{xxx}	224.142
β_{xxy}	223.0142
β_{xyy}	152.4782
β_{yyy}	356.2156
β_{xxz}	120.0124
β_{xyz}	-71.4309
β_{yyz}	-132.2451
β_{xzz}	24.8258
β_{yzz}	198.2541
β_{zzz}	-38.2541
β_{tot}	$7.57352\text{E-}30 \times 10^{-30}$

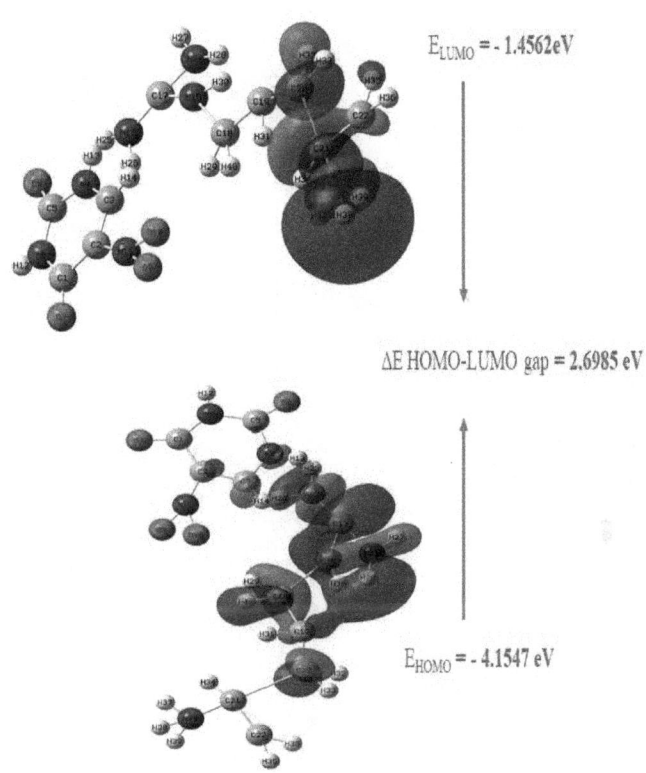

Fig. 4.5 HOMO – LUMO plot of LA5N at B3LYP/6-311++G (d, p)

Table: 4.5: Calculated electronic and energies of LA5N using B3LYP/6-311++G (d, p) level

Electronic parameter	Calculated values
E_{HOMO} (eV)	-4.1547
E_{LUMO} (eV)	-1.4562
$\Delta E_{HOMO} - E_{LUMO}$ (eV)	2.6985
Ionization Potential (IP)	4.1547
Electron Affinity (EA)	1.4562
Chemical Hardness (η)	1.3492
Softness (s)	0.3705
Electronegativity (χ)	2.8054
Chemical Potential (μ)	-2.8054
Electrophilicity index (ω)	2.9166

reactivity and low kinetic stability (H. Chermette et al 1999). The calculated energy values for title compound in the gas phase are; $E_{HOMO} = -4.1547$ eV, $E_{LUMO} = -1.4562$ eV.

The LA5N compound has a low chemical softness and chemical potential of -2805. The narrow energy gap of LA5N as seen in Fig.4.5. The low value of softness theoretically proves the reduced toxicity of the title molecule while the significantly high electrophilicity index is a biological activity descriptor.

4.3.7 THERMODYNAMICAL PROPERTIES

The standard thermodynamic functions can be used as reference thermodynamic values to calculate the changes of entropies (ΔS), changes of enthalpies (ΔH) and changes of Heat capacity (ΔC) of the reaction. The dipole moment and its principal inertial axes strongly depend upon the molecular conformation. From Table 4.6, It can be observed that these thermodynamic functions are increasing with temperature ranging from 100 to 1000 K due to the fact that the molecular vibrational intensities increase with temperature (Pathak et al 2015). The following are the associated fitting conditions, with their relationship charts shown in Fig. 4.6, 4.7 and 4.8. The thermodynamic data (enthalpy, entropy, heat capacity) are non-linear function of temperature. The polynomial equations arrived from thermo.pl for Capsaicin has been presented below. They express the temperature dependance of enthalpy, entropy, heat capacity of capsaicin with best fitting factors. The correlation equations between heat capacities, entropies, enthalpy changes and temperatures were fitted by quadratic formulas, and the corresponding fitting factors (R^2) for these thermodynamic properties are 0.9974, 0.9751 and 0.9896, respectively.

$$S^0{}_m = 0.530x^2 + 29.884T - 25.819 \times 10^{-5} T^2 \ (R^2 = 0.9751)$$

$$Cp^0{}_m = 0.474x^2 + 32.92T - 53.635x \ 10^{-5} T^2 \ (R^2 = 0.9974)$$

$$\Delta H^0{}_m = -0.288x^2 + 20.432T + 92.267x \ 10^{-5} T^2 (R^2 = 0.9896)$$

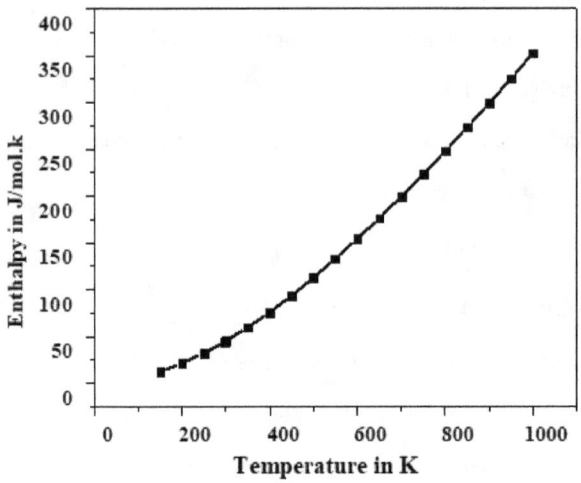

Fig. 4.6 Variation of Enthalpy with temperature

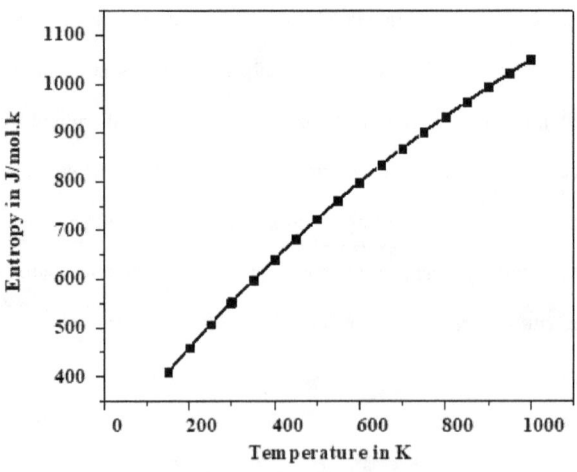

Fig. 4.7 Variation of Entropy with temperature

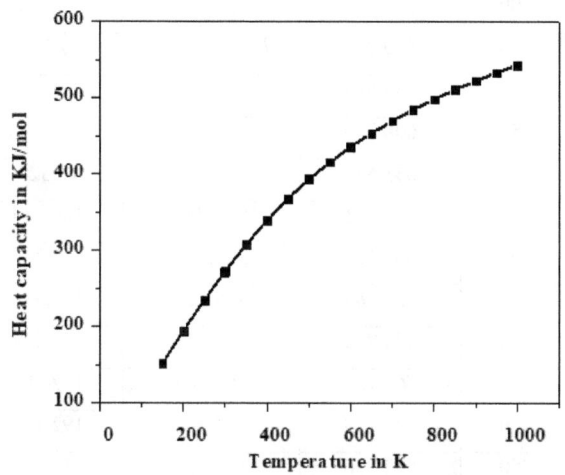

Fig. 4.8 Variation of heat capacity with temperature

Table 4.6 Thermodynamic properties at different temperatures by B3LYP level for LA5N

T (K)	S (J/mol.K)	Cp (J/mol.K)	ΔH (kJ/mol)
100	329.823	194.158	16.32
150	396.255	147.254	17.9
200	442.751	178.241	22.475
250	482.246	195.256	29.365
300	525.167	220.356	39.354
350	549.241	239.856	52.368
400	587.983	274.154	63.253
450	608.245	249.325	79.354
500	643.585	303.321	96.325
550	657.391	315.256	109.547
600	698.127	326.145	125.256
650	725.583	335.222	142.586
700	749.222	349.258	156.324
750	769.757	359.365	172.589
800	781.429	368.254	192.142
850	802.256	370.254	212.253
900	820.291	375.214	235.256
950	833.583	380.236	249.369
1000	885.179	398.258	270.365

4.3.8 OPTICAL STUDIES

The experimental UV-Visible spectrum of a LA5N molecule was recorded by using methanol as solvent and it is compared with simulated spectrum calculated using TD-DFT/B3LYP method with 6-311++G (d, p) basis. Time-dependent DFT can predict the excited state properties like band gap energy. The maximum absorption (λ max) of our title molecule is estimated and also computed the band gap energy. The maximum peak λ max at 431nm showing a good match with measured experimental data of 428 nm. The very strong band observed at 431nm. The band gap energy was estimated using the formula, $E = hc/\lambda$. Here h and c are constant; λ is the cut off wavelength. The optical band gab energy value is found to be 2.94 eV is shown in Fig. 4.10. The spectrum reveals that the transmittance of 1 mm thickness LA5N crystal is ~ 75% in entire range of spectrum, further it is observed that the transmission of crystal sharply falls to minimum value in lower wavelength region. Figs 4.12- 4.14 illustrate the variance of reflectance (R) with the wavelength and the difference of refractive index (n) and the extinction coefficient (k) with photon energy of the LA5N crystal.

4.3.9 MULLIKEN POPULATION ANALYSIS

The natural population analysis of the title molecule (NPA) is obtained by Mulliken population analysis (MPA) using B3LYP 6–311++G (d, p) method. The charge and multiplicity are varied in order to compare the variation in the Mulliken charges in each case. Mulliken atomic charge calculation helps in understanding the chemical potential and ionization potential. Atomic charge affects dipole moment, polarizability, electronic structure and other molecular properties of the system. Charge distribution of L- Argininium 5-Nitrouracilte shows all the hydrogen atoms are positively charged, while the magnitudes of the atomic charges on carbon atoms for the L- Argininium5-Nitrouracilte compound. (LA5N) compound as seen in Fig.4.15. were observed to be both positive and negative ranging from −3.7578 to 3.7581. When we compare the atomic charge for the all atoms, we note that the maximum atomic charge is obtained for C2. This is owing to negatively charged carbon C3 atom attached. Moreover, the oxygen atom exhibits the large negative charge. The presence of negative charge on the oxygen atom and net positive charge on the hydrogen atom propose the establishment of intermolecular interaction and consequently. Mulliken atomic charges of LA5N single crystal was illustrated in Table 4.7.

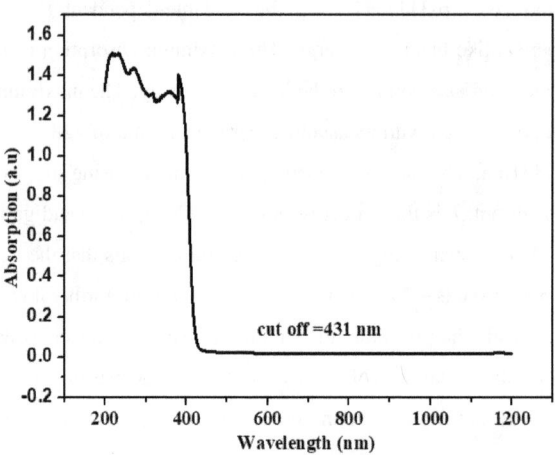

Fig. 4.9 UV- Vis absorption spectrum of LA5N crystal

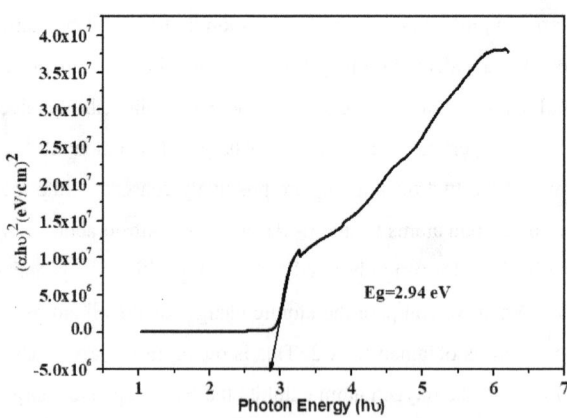

Fig. 4.10 UV- Vis bandgap plot of LA5N crystal

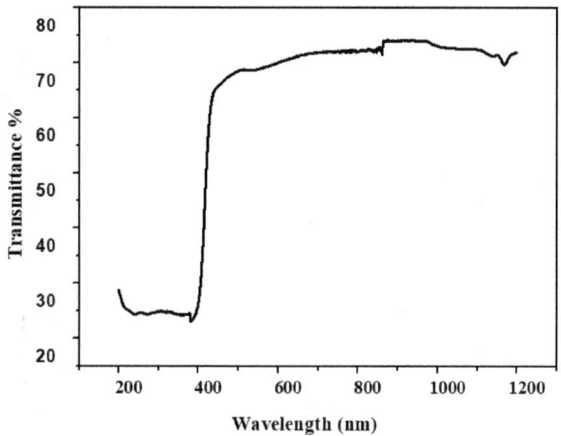

Fig. 4. 11 UV-Vis transmission spectrum of LA5N crystal

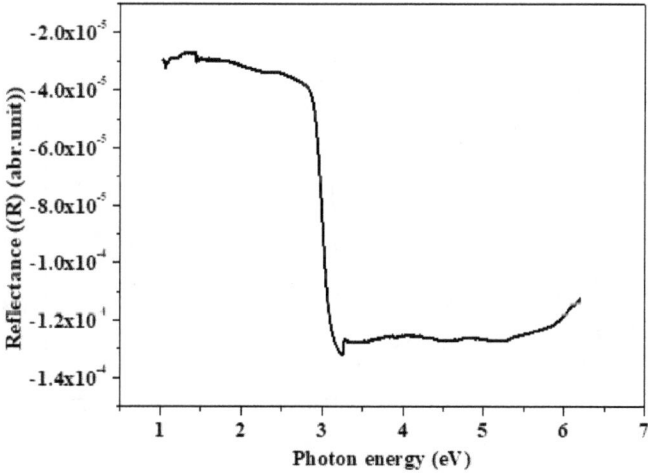

Fig. 4. 12 UV- Vis reflective index of LA5N crystal

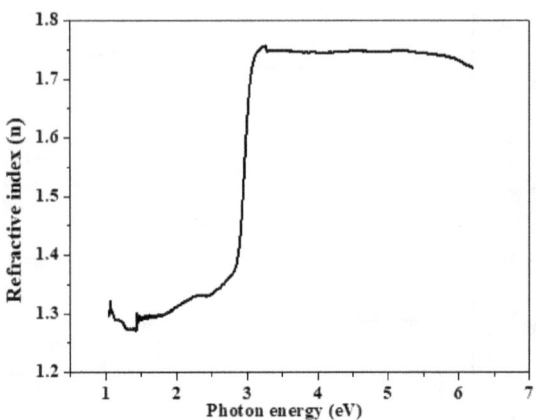

Fig. 4. 13 UV- Vis refractive index of LA5N crystal

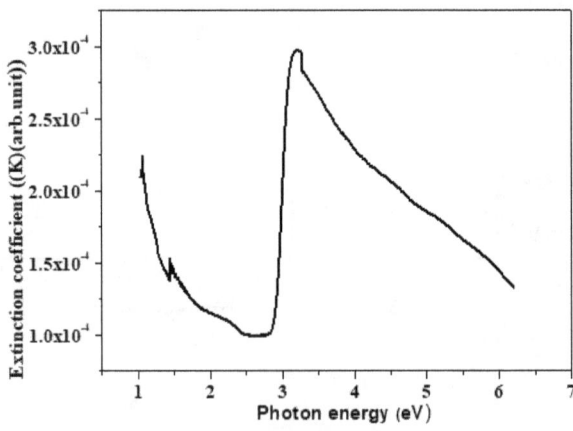

Fig. 4. 14 UV-Vis extinction co-efficient of LA5N crystal

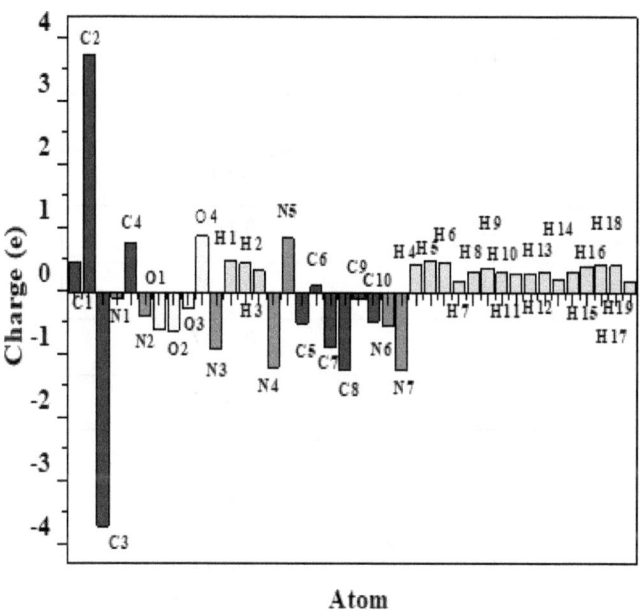

Fig. 4.15 Mulliken atomic charges of LA5N single crystal

Table 4.7: Mulliken atomic charges of LA5N single crystal

Atoms	Mulliken atomic charge
C1	-0.7152
H1	0.1232
H2	0.0967
C2	-0.0808
H3	0.2472
N1	-0.8657
H4	0.4137
H5	0.4297
H6	0.4137
C3	-0.4735
H7	0.1667
H8	0.2472
C4	0.1671
H9	0.2311
C5	-0.2368
H10	0.2361
H11	0.2953
N2	-0.8386
H12	0.4192
C6	1.0321
N3	-0.8601

H13	0.4192
H14	0.4081
N4	-0.8601
H15	0.4032
H16	0.4081
C7	0.4624
C8	0.2903
C9	0.3064
C10	0.2848
H17	0.2096
O1	-0.5697
O2	-0.6289
O3	-0.6400
O4	-0.7208

4.3.10 NATURAL BONDING ORBITAL (NBO) ANALYSIS

Natural bonding orbital (NBO) examination is effective techniques to analysing inter and interamolecular interactions in nuclear bonding, as well as convenient way to observe charge transport between molecular system. The stabilisation energy ($E^{(2)}$), which is determined by the second order perturbation theorem, is obtained as

$$E^{(2)} = = q_i \frac{(i,)^2}{\Sigma_j - \Sigma_i}$$

The donor orbital occupancy is q_i, the diagonal elements are E_i and E_j and Fock matrix elements are $F(i,j)$.

The DFT (B3LYP) / 6-311++ G (d, p) level, NBO was performed on the LA5N to reveal the delocalization electron density, rehybridization, inter and interamolecular interaction that occurs in the compound. For LA5N compound, the energy difference between the donor (i) and acceptor (j) was calculated and tabulated in table 4.8.

The high satabilization energy obtained from σ^* (C2-C3) to π^* (O9-N11) with an energy value of 10.50 Kcal/mol. The stabilization energy of n1 (N6) to σ^* (C1) lone pair electron is 58.93 Kcal/mol. With stabilisation energies of 10.11 and 10.50 Kcal/mol, the main interaction with high stability of the LA5N molecule was obtained from the donor orbitals of σ (N4) and σ (N6) to acceptor orbitals of π (C2-C3) and π^* (C1-O7) correspondingly.

4.3.11 MICROHARDNESS TEST

The microhardness analysis was accomplished by Leitz wetzler Vickers microhardness tester to investigate its mechanical stability. Microhardness is one of the best methods to analyze the mechanical strength of the material. The mechanical strength of the material plays a vital role in the manufacture of the device. The hardness of a material is a measure of resistance against the motion of dislocation, deformation, or damage under an applied stress. LA5N crystal with a clear surface has been subjected to Vickers indentation test at room temperature. Loads of different weights such as 25, 50, 100 g were applied and the indentation time was maintained as 10s for all loads applied. Beyond 100g of the applied load, crack initiation and fragmentation were observed. So, the hardness test could not go beyond this load. Fig.4.16 shows the plot between the loads applied by P and the corresponding hardness number HV. The graph shows linear or nonlinear behaviour primarily corresponding to the

Table 4.8: Second order perturbation theory analysis of Fock Matrix in NBO for LA5N with 6-311++G (d, p) basis set

Donor (i)	Type	Acceptor (i)	Type	$E(2)^a$ (k cal/mol)	$E(j)-E(i)^b$ (a.u)	$F(I,j)^c$ (a.u)
C1-O7	π	C1	σ^*	19.82	0.92	0.121
C2-O7	π	C1	σ^*	10.11	1.04	0.092
C2-C3	σ^*	O9-N11	π^*	10.50	1.06	0.094
C2-C3	π	C1-O7	π^*	16.61	0.29	0.066
O9-O10	σ	C2-N11	σ^*	12.70	1.72	0.132
O8	σ	C5	σ^*	10.11	19.59	0.399
N4	σ	C2-C3	π^*	42.52	0.27	0.096
N4	σ	C5-O8	π^*	55.03	0.27	0.109
N6	σ	C1-O7	π^*	63.13	0.26	0.115
N6	n1	C1	σ^*	58.93	0.26	0.111
O7	π	C1	σ^*	20.75	0.66	0.106
O7	σ	C1-N6	σ^*	31.58	0.65	0.130
C5	σ	O8	n1	18.87	1.61	0.156
O8	π	N4-C5	σ^*	26.42	0.65	0.119
N11	π^*	C2-C3	σ	23.80	0.29	0.077

cleavage plane of the crystal. The good hardness number elevation in relation to the load describes the material with high mechanical strength. The hardness of the material is guaranteed from the plot between log P and log d in Fig.4.17. The value of the Meyer index and also known as the work hardening coefficient 'n' is calculated from the graph slope between log P and log d shown in the Fig. 4.17. The value of the slope (n) was found to be 1.15. Hence, it is concluded that LA5N is a hard material category.

4.3.12 THERMAL ANALYSIS

The thermal behaviour of the LA5N was initially investigated by means of Thermogrammetric Analysis (TG) and Differential Thermal Analysis (DTA) using the LA5N sample of 7 mg. A thermal analyzer was used in the nitrogen atmosphere at a heating rate of 20°C/min. The thermogram obtained from analyzes of TG and DTA recorded at 20-500 °C is shown in Fig. 4.18. The TGA spectrum indicates that up to 220 °C there is no weight loss which confirms the lack of crystallization water in the title compound. Two exothermic peaks are shown in the differential thermal analysis. It implies that the material undergoes an exothermic transition at 244°C which promptly indicates that the first stage of the material's decomposition well matches the TGA spectrum. The second endothermic peak at 272°C can be due to the compound's decomposition and volatilization. Thereafter no sharp peak has been observed, confirming that the material is thermally stable up to 150 °C. The sharpness of the peaks shows good degree of crystallinity of the grown crystal.

4.3.13 DIELECTRIC STUDIES

The dielectric constant and dielectric loss as a function of temperature and frequency are shown in Fig 4.19 and 4.20. From Fig 4.19. it shows the value of dielectric constant is found to increase with temperature and it becomes independent of frequency at higher frequency region. The decrease in dielectric constant of LA5N crystal at low frequencies may be attributed to the contribution of the electronic, ionic, orientation and space charge polarizations which depend on the frequencies (C. Balarew et al 1984). As the frequency increases, it is found that the dielectric constant decreases exponentially and it attains a lower value. The low value of dielectric loss at high frequencies suggests that the sample possess enhanced optical quality with lesser defects and this parameter is of vital importance for NLO applications.

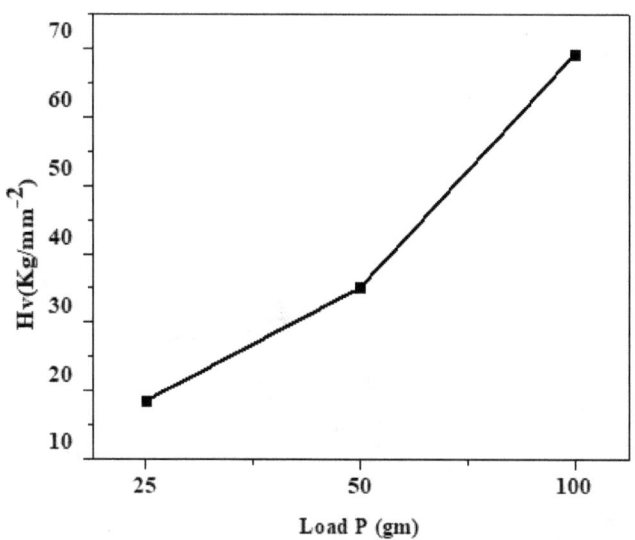

Fig. 4.16 Vickers hardness number Vs applied load of LA5N

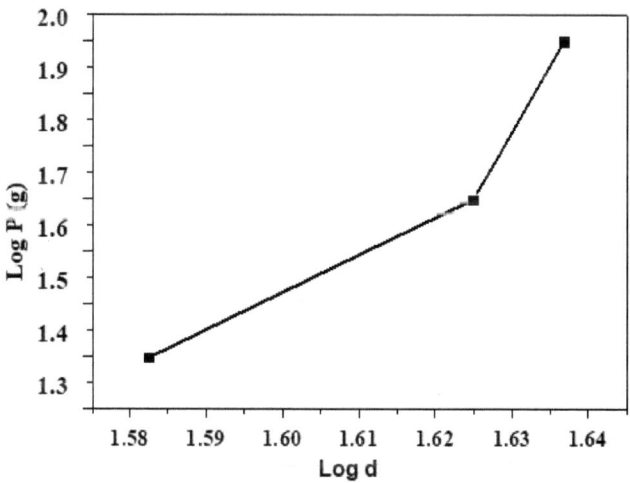

Fig. 4.17 log P Vs log d plot of LA5N

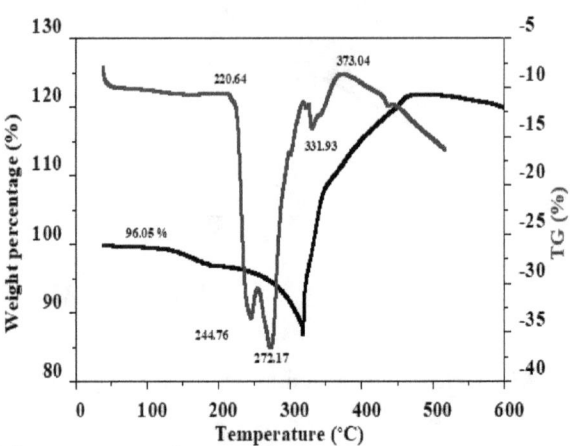

Fig. 4.18 TG-DTA plot of LA5N crystal

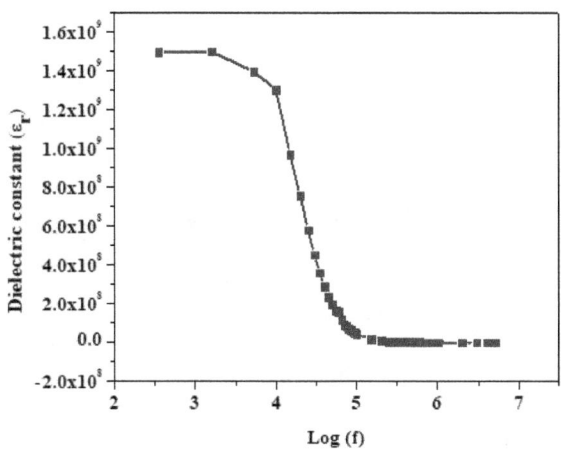

Fig. 4.19 Variation of dielectric constant of LA5N

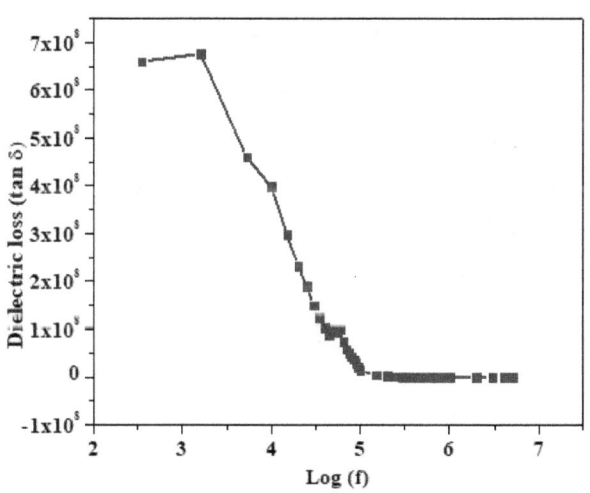

Fig. 4.20 Variation of dielectric loss of LA5N

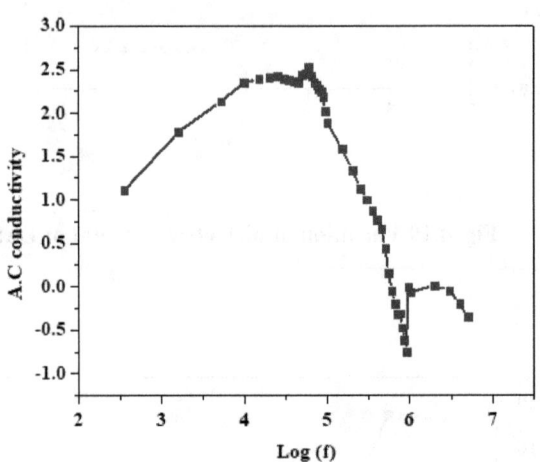

Fig. 4.21 Frequency dependence of AC Conductivity

4.3.14 PHOTOLUMINESCENCE ANALYSIS

Photoluminescence spectral analysis is one of the effective tools for providing relatively direct information on the physical properties of materials at the molecular level, including shallow and deep defects and gap states. In order to identify the photoluminescence characteristics of LA5N, the grown crystals were excited at 294 nm. The emission spectrum is recorded at room temperature in the range of 200–700 nm which is shown in Fig 4.22. One high intensity sharp emission peak centered at 492 nm and a broad violet band at 422 nm. The highest peak is observed at 492 nm in the PL spectrum, which indicates the bluish-violet emission in the visible region and also high purity of the grown crystal. These results show the grown material is sorely applicable for LED devices.

4.3.15 MOLECULAR ELECTROSTATIC POTENTIAL

The molecular electrostatic potential (MEP) is related to the electronic density and is a very useful descriptor for reactive sites for electrophilic and nucleophilic reactions as well as hydrogen bonding interactions (F.J. Luque et al 2000 and N. Okulik et al 2005). Electrostatic potential map shown in Fig 4.23 illustrates the charge distributions of the molecule three dimensionally. As it can be seen from this figure, the different values of the electrostatic potential at the surface are represented by different colours; red represents regions of most electronegative electrostatic potential, blue represents regions of the most positive electrostatic potential and green represents region of zero potential. The negative electrostatic potentials are shown in red (Oxygen atoms) and yellow, slightly electron rich region (Nitrogen atom) the intensity of which is proportional to the absolute value of the potential energy, while green indicates surface areas where the potentials are close to zero. The colour code of these maps is in the range between $-7.112e^{-2}$ (deepest red) to $+7.112e^{-2}$ (deepest blue) in title compound, where blue indicates the strongest attraction, minimal concentration of electrons and most positive electrostatic potential whereas red indicates the strongest repulsion, high density concentration of electrons and most electronegative electrostatic potential.

4.3.16 SECOND HARMONIC GENERATION (SHG)

The second harmonic generation (SHG) activity of the grown crystal was examined by Kurtz and Perry technique. It is an important tool to evaluate the conversion efficiency of NLO materials.

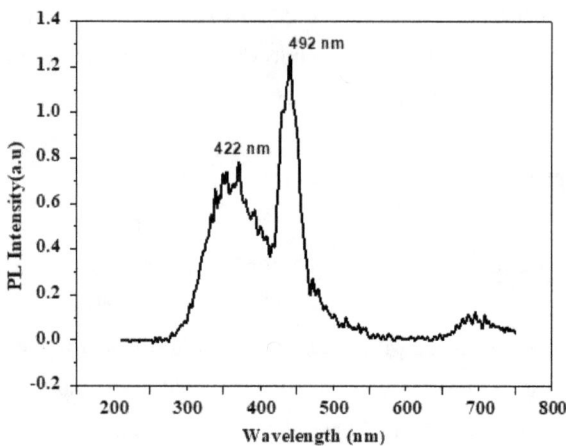

Fig. 4.22. Fluorescence emission spectra of LA5N crystal

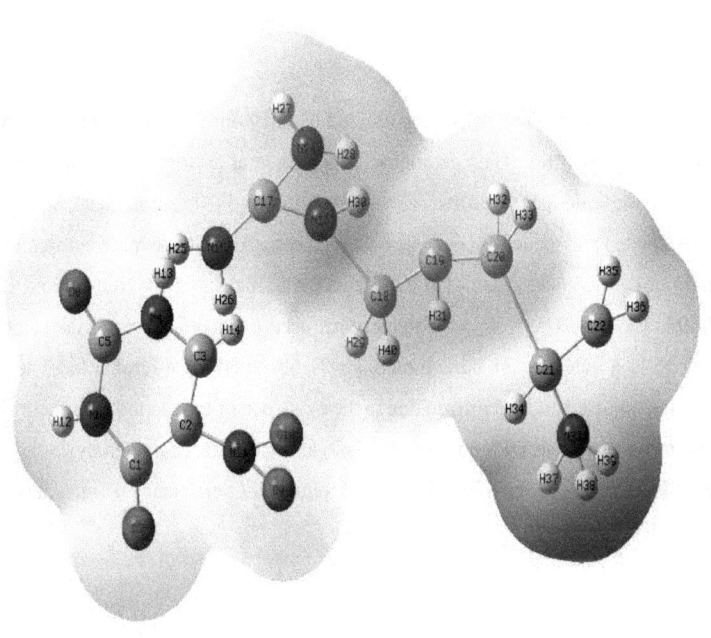

Fig. 4.23. Molecular Electrostatic Potential (MEP) of LA5N crystal

The LA5N single crystal is approximately equal in second order harmonic efficiency when compared with the standard Potassium dihydrogen phosphate (KDP). The LA5N single crystal is 1 time greater in second order efficiency (20 mV) when compared with the standard Potassium dihydrogen phosphate (KDP) material (20 mV). From this analysis to conclude that the LA5N crystal would be an optional material in SHG device fabrications.

4.4 CONCLUSION

The 6-311++G (d, p) DFT method is utilized to carry out a full vibrational analysis in the compound of the title. The geometric experimental settings generated a good approximation of the calculated results. The L-Argininium 5-Nitrouracilte (LA5N) FT-IR spectrum was recorded and the numbers wave observed were found to be quite compatible with the calculated ones. The energy gap of the molecule was measured by HOMO and LUMO studies as 2.6985 eV. The UV-Vis spectrometer was used to record the material's electronic spectrum. LA5N is a category of hard material was observed by microhardness test. The substance was decomposed and volatilized due to the existence of the second endothermic peak of 272°C. The molecule's natural bind orbital (NBO) and mulliken charge analysis shows that there is intermolecular charge transfer. SHG study found NLO efficiency 1 times larger than the conventional KDP material.

CHAPTER V

EXPERIMENTAL AND COMPUTATIONAL STUDIES ON L-GLYCINIUM 5-NITROURACILATE (LGY5N) SINGLE CRYSTAL

5.1 INTRODUCTION

The search for new advanced materials is an important area of contemporary research in numerous disciplines of science and development of many new technologies. The nonlinear optical (NLO) crystal has become of great research interest and importance in the recent years for the fabrication of devices used in the field of telecommunication, optical signal processing, optical switching and photonics (S. Arulmani et al 2019). Now a day, various growth methods and apparatus have been continuously developed to improve the quality and growth rate. Compared to the other techniques, the slow evaporation technique is mostly used in several types of crystals. Organic crystals in terms of NLO properties possess advantages when compared with inorganic counterparts [S. Senthil et al 2012]. Organic materials allow their fine tuning of their chemical structure and properties for the desired NLO properties (S. Arulmani et al 2018). The adaptable nonlinear optical frequency conversion materials are vital importance of optical modulation, optical switching, optical logic, optical storage, computing and optical information process (M. Amudha et al 2017). Single crystals formed by amino acids such as L-arginine, L-histidine, etc., are found to be promising nonlinear optical (NLO) materials (C. Preema et al 2005). Organic materials draw more interest because of their superior performances involving fairly high NLO coefficient and fast response than their inorganic counterparts (Datta et al 2003). The authors of (K Ambujam et al 2007) obtained the salt L-histidinium 5-nitrouracilate and the presence of amino acid L-histidine chiral molecule inevitably leads to the formation of a non-centrosymmetric structure. So, we tried to obtain a L-Glycinium 5-Nitrouracilate (LGY5N) single crystal. So, it crystallizes in the polar space group $P2_1$ and displays highest non-linear optical properties among the L-Glycinium salts. According to (S.R. Domingos et al 2012 and M. Fleck et al 2014), L-Glycinium (as well as other amino acids, having a basic side chain) can form salts not only with singly, but also with a doubly charged cation. In the present work, we show the results of the structural investigation of the new monoclinic form of single crystal. In addition, the influence of the crystal structure on the nonlinear optical properties has been studied.

5.2 SYNTHESIS AND GROWTH

The compound of L-Glycinium 5-Nitrouracilate (LGY5N) Single crystal were grown from supersaturated solution by slow evaporation technique. The substance of L-Glycinium- $C_2H_5NO_2$ (Sigma–Aldrich, 99%) and 5-Nitrouracil - $C_4H_3N_3O_4$ (Sigma–Aldrich, 99%) were taken in equimolar ratio (1:1). The solution was mixed together to dissolve in water until to attain the supersaturated solution. Then, the highly dissolved salt solution is filtered by whattman filter paper and poured in a beaker. The beaker is covered with a silver foil paper with a few small holes for slow evaporation of solvents. The apparatus is placed undisturbed till the seeds of sufficient size of LGY5N are obtained in the beaker. The LGY5N single crystal was harvested within 40-45 days from the mother solution. The photograph of LGY5N single crystal is shown in Fig.5.1.

5.3 RESULTS AND DISCUSSION
5.3.1 SINGLE CRYSTAL X-RAY DIFFRACTION

The single crystal x-ray diffraction (SCXRD) analysis of the grown LGY5N single crystal were studied by using the Enraf Nonius CAD4-MV31 single crystal with MoKα (λ=0.717073 Å) radiation. It is used to confirms the crystal structure and lattice parameters value etc., The obtained lattice parameter values are a=7.8254 Å, b=8.2541 Å, c= 14.528 Å, α = 90°, β= 93.16°, γ=90°. The Structure of LGY5N single crystal was monoclinic and the space group was $P2_1$. The crystal parameters were mentioned in table 5.1.

5.3.2 COMPUTATIONAL DETAILS

Density functional theory (DFT) is an effective tool for quantum chemical computation of molecules. All calculations in the present work were performed using the Gaussian 09 W (H.A. Petrosyan et al 2005) program package on a personal computer. Geometrical parameters and vibrational wavenumbers were computed by optimizing the geometry of the molecule using the B3LYP method with 6-311++G (d, p) as the basis set. The energy distribution from HOMO to LUMO was calculated using the Gaussian software program which served as a precursor to theoretically attain the band gap of the title molecule.

Fig. 5.1 Photograph of as grown LGY5N single crystal

Table 5.1: Crystal parameters of LGY5N single crystal

Empirical Formula	$C_6H_8N_4O_5$	
Formula weight	256.27 g/mol	
Crystal system	Monoclinic	
Space group	$P2_1$	
Unit cell dimensions	a = 7.8254 Å	α = 90°(5)
	b = 8.2541 Å	β = 93.16°(5)
	c = 14.528 Å	γ = 90°(5)
Cell volume	1746.42 $Å^3$	
Absorption coefficient	0.142 mm^{-1}	
F(000)	634	
Crystal size	0.3 x 0.50 x 0.40 mm^3	
Theta range for data collection	3.213 to 24.899°.	
Limiting indices	-6<=h<=6, -16<=k<=16, -18<=l<=18	
Reflections collected/unique	1784/ 2074	
Completeness to theta = 23.92	95.4%	
Refinement method	Full-matrix least-squares on F^2	
Data/restraints/parameters	2214 / 1 / 176	
Goodness-of-fit on F2	1.147	
Final R indices [I>2sigma(I)] 80	R1 = 0.0452, wR2 = 0.0571	
R indices (all data)	R1 = 0.0348, wR2 = 0.0177	

5.3.3 MOLECULAR GEOMETRY

The optimized structure of the molecule is obtained from Gaussian 09 Gauss view program. The bond lengths and bond angles of LGY5N are compared with experimentally observed geometrical parameters. This molecule has three C-C bonds, six C-N bonds, four C-O bonds, two N-O bonds, and four N-H bonds. The bond distance between C3-C4 and C5-N5 has been calculated to be 1.392 Å and 1.385 Å respectively. The lengths of other bonds like; N6-C2 and C2-O7 present in LGY5N are also, in well correlation with experimental values. The molecular structure along with numbering of atoms of is shown in the Fig. 5.2. The most optimized structural parameters were calculated and were depicted in Table 5.2.

5.3.4 VIBRATIONAL ASSIGNMENTS

The title compound consists of N = 25 atoms and hence has 66 modes of vibrations (3N-6) apart from three translational and three rotational degrees of freedom. The theoretical spectrum obtained from Gaussian 09 W using B3LYP/6-311++G (d, p) basis set and the experimental spectrum were compared in Fig.5.3. and 5.4 respectively. The potential energy distribution for each normal mode among the symmetry coordinates of the molecules is calculated. The vibrational band at 3025 cm^{-1} was assigned to the vibration of N-H band. The band at 1698 cm^{-1} is due to bending vibrational mode of O-H group. The broad band around ~3246 cm^{-1} was due to the O-H vibrations. Bands in the region from 3100 to 3700 cm^{-1} are usually due to various O-H and N-H stretching vibration. The bonded O-H groups usually give rise to a broader band than N-H groups. In nitro compounds the anti-symmetric N-O stretching vibrations are located in the region 1565 cm^{-1}. The peak at 1502 cm^{-1} is due to C-O stretching vibration. It shows that the theoretical values are in good agreement with experimental data.

VIBRATIONS OF CN GROUP

Assignment of the C-N stretching wavenumber is a rather difficult task since there are problems in identifying these wavenumbers from other vibrations. The C-N stretching vibration is moderately to strongly active in the range 1230–1310 cm^{-1}. In this study, we have assigned the FT-IR band observed at 1231 cm^{-1} to the C-N stretching mode. The theoretically predicted value at 1211 cm^{-1} by the B3LYP method coincides well with the observed wavenumber. The C-N in-plane bending vibration usually occurs at 527 cm^{-1} (V. Krishnakumar et al 2003). The DFT computation predicts this vibrational mode at 448 cm^{-1}.

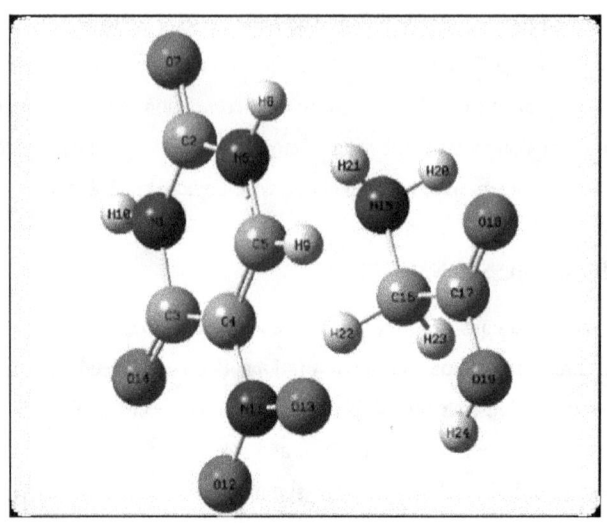

Fig. 5.2 Atomic numbering system adapted for ab initio computations of LGY5N molecule

Table 5.2: Selected bond lengths of LGY5N molecule

Parameters	Bond length (Å)	
	Observed Bond Length	Calculated Bond Length
C3-C4	1.392	1.321
C4-C5	1.396	1.402
C5-N5	1.385	1.395
N6-C2	1.381	1.354
C5-H9	0.910	1.010
C2-O7	1.230	1.210
C3-O14	1.236	1.021
C4-N11	1.417	1.398
N11-O12	1.245	1.354
N11-O13	1.230	1.100
O13-H23	0.916	0.956
N15-C16	1.487	1.354
C16-C17	1.509	1.425
C16-H23	0.970	0.954
C17-O18	1.299	1.112
O19-H24	0.980	0.963
N1-H10	0.910	0.910
N15-H20	0.910	0.910
N15-H21	0.970	0.845

Table 5.3: Selected bond angles of LGY5N molecule

Parameters	Bond Angles in (°)	
	Observed Bond Angle	Calculated Bond Angle
C3-C4-C5	122.101	123.541
C2-N6-C5	111.281	110.231
O7-C2-N6	116.465	115.201
C5-C4-N11	152.177	153.214
N6-C5-H9	128.032	130.236
N1-C3-O14	121.203	120.254
N15-C16-C17	176.459	177.548
O18-C17-O19	91.776	90.248
O13-H23-C16	161.546	162.965
C4-C3-O14	133.477	132.582
C3-N1-H10	124.664	125.652
O7-C2-N1	127.885	125.365
H21-415-H20	145.614	146.236
H23-C15-N15	91.671	90.326
N1-C2-N6	115.649	114.201
O7-C2-N5	116.465	115.236
C2-N6-H8	167.632	165.247
C5-N6-H8	81.086	80.236
N6-C5-C4	126.851	125.325
C4-C3-O14	133.477	132.254
C4-N11-O12	132.716	130.236
O19-C17-O18	91.776	90.254
N11-O13-H23	163.463	161.254
C16-C17-O19	1662.229	165.845
C4-N11-O12	132.716	133.254

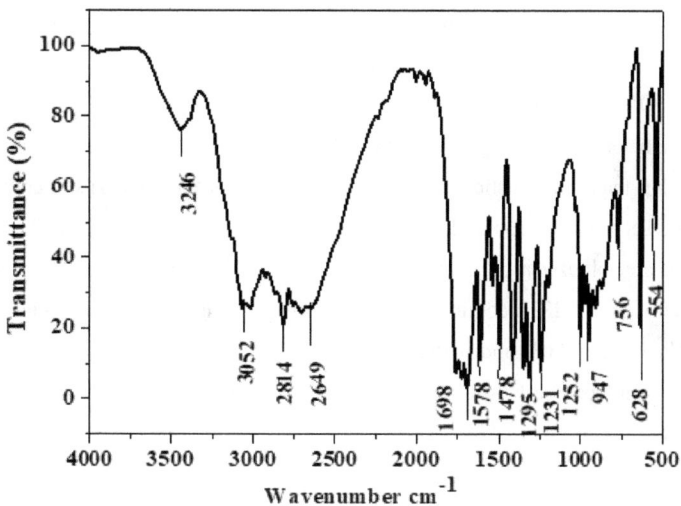

Fig. 5.3 Experimentally obtained FT-IR spectrum of LGY5N

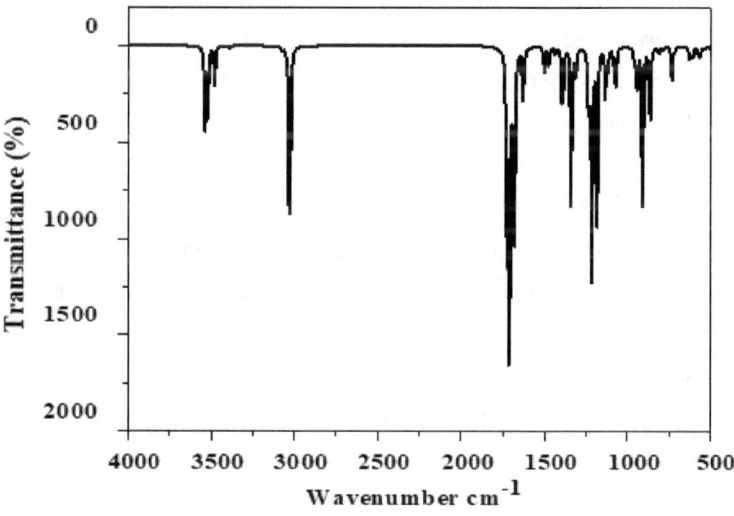

Fig. 5.4 Theoretically simulated FT-IR spectrum of LGY5N

The C-N out-of-plane bending vibration is generally found at 326 cm^{-1} (J. Swaminathan et al 2009). In the present study, the C-N out-of-plane bending vibration is assigned to a weak band at 325 cm^{-1} and shows a good agreement with the expected value.

VIBRATIONS OF CH GROUP

The heterocyclic compounds have C-H stretching vibration in the region 3200-3000 cm^{-1} which is the characteristic region for the ready identification of C-H stretching vibrations (G. Varsanyi et al 1973). The C-H stretching modes generally appear in strong Raman intensity regions and are polarized highly. In the title molecule, C-H stretching vibrations were observed at 3052 cm^{-1} in FT-IR spectrum. The peak corresponding to C-H stretching vibration is observed in the range 3033 and 3019 cm^{-1} by B3LYP/6–311++G (d, p) method shows excellent agreement with recorded FT-IR spectral values. The C-H stretching vibrations are very weak in intensity.

VIBRATIONS OF CC GROUP

The ring stretching vibrations are very much important in the spectrum of benzene ring and its derivatives, highly characteristic of aromatic ring itself. The ring C-C and C-C stretching vibrations, known as semi-circular stretching usually occurs at the region 1390-1650 cm^{-1}. The observed bands are at 1686, 1578, 1231, 947 and 756 cm^{-1}. Theoretically computed vibrations are present at the range of 1672, 1548, 1254 and 781 cm^{-1} respectively. These values are in good agreement with the measured values.

NITRO GROUP VIBRATION

The NO_2 stretching vibrations are very useful group vibrations because of their spectral position and strong intensity. The NO_2 asymmetrical stretching vibrations in nitro-alkanes occur in the range 1580 -1530 cm^{-1} and the symmetric vibration lie in the range 1380–1360 cm^{-1}, the asymmetrical stretching being the stronger than the symmetrical stretching. In aromatic compounds, the NO_2 stretching bands shift down to slightly lower wavenumbers in the range 1540-1500 cm^{-1} and 1370-1330 cm^{-1} (B. Stuart et al 2010). In the LGY5N molecule there are two nitro groups at meta position with respect to each other which are attached to the benzene ring. The wavenumbers calculated at 1587, 1583 cm^{-1} and 1368, 1360 cm^{-1} correspond to the asymmetric and symmetric stretching of nitro group and are assigned well in the spectra. While the hydrogen bonding has very insignificant effect on asymmetric stretching it shows appreciable effect on the symmetric NO_2 vibration (Xitao Liu,et al 2014), hence the deviation in these vibration wavenumbers from the experimental values is justified. The modes calculated at 916, 912 cm^{-1} is a mixed mode having contributions from NO_2 scissoring.

VIBRATIONS OF OH⁻ GROUP

The O-H stretching vibrations are extremely sensitive to hydrogen bonding. The nonhydrogen bonded (or) free hydrogen group absorbs strongly in the region 3500-3502 cm^{-1}, whereas the existence of intermolecular hydrogen bond formation can lower the O-H stretching frequency to the 3550-3200 cm^{-1} region with increase in intensity and breath (K. Bahgat et al 2007, B. Smith et al 1999, and R. M. Silverstein et al 2003). The O-H out-of-plane bending vibrations of hydroxyl group are expected in the regions 517-710 cm^{-1} for free and associated O-H groups (G. Varsanyi et al 1974).

VIBRATIONS OF AMMONIA

Vibrational band due to the N-H stretching vibration appears in the region 3352 ± 60 cm^{-1} as a broad strong band (S. Gunasekaran et al 2006). The frequencies of the amino group appear at around 3400-3300 cm^{-1} for the N-H stretching, 1700-1600 cm^{-1} for the scissoring and 1150-900 cm^{-1} for the rocking deformations (G. Socrates et al 2001). The position of absorption in this region depends on the degree of hydrogen bonding and hence on the physical state of the sample. The spectral lines assigned to N-H stretching vibrations have shifted to higher region in the present system. This clearly indicates that the stretching of N-H bond depends on protonation and shifts the frequency to the higher region.

VIBRATIONS OF C=O GROUP

The ring carbon–carbon stretching vibrations occur in the range of 1610-1410 cm^{-1} and in general the bands are of variable intensity. In the present work, the wavenumbers observed in the FT-IR spectrum at 1580, 1325 and 1233 cm^{-1} have been assigned to C-C stretching vibrations. The theoretically computed values by the B3LYP/6- 311++G (d, p) method at 1611, 1300 and 1224 cm^{-1} show an excellent agreement with experimental data. The in-plane bending modes are at higher wavenumbers than those of out- of plane vibration. The bands at 756, 743 and 628 cm^{-1} in the infrared are assigned to the C-C-C in-plane bending modes. The predicted wavenumbers of C-C-C in-plane bending vibrations calculated at 741, 734 and 651 cm^{-1} are in good agreement with the measured values.

5.3.5 HYPERPOLARIZABILITY

In the domain of NLO, the incident electromagnetic fields altered into higher order fields by means of phase, frequency and amplitude when the electromagnetic fields move to various medium. Due to an induced dipole moment (μ), the polarization of the molecule is estimated systematically. To determine the dipolar interaction by the whole of the external electric field, the Taylor series expansion should be employed in the components of an electric field in case of poor polarization circumstance. The finite field approach is the basic principle to obtain the molecular first polarizability (β_0) and its associates (β, α_0 and $\Delta\alpha$) of LGY5N. In this calculation, Becke three parameter hybrid exchange functional and the Lee-Yang-Parr correlation. The energy of a system is established due to the external electric field. For a 3x3x3 matrix, 3rd rank tensor (first hyperpolarizability) should be required. The Kleinman symmetry is used to diminish the components from twenty-seven to ten (D.A. Kleinman et al 1977). The energy coefficients in the Taylor series expansion described the dipole moment components, polarizability and the first order hyperpolarizabilities. The elements of the electric dipole moment (μ), Polarizability (α) and molecular first polarizability (β) values of LGY5N have been calculated and presented in table 5.4. The determined first order hyperpolarizability of LGY5N is 7.9010X 10^{-30} e.s.u. The value of β is exceeding the molecular first polarizability value of urea as 7 times. The static dipole moment (μ) is 1.1754 Debye, which has been compared with the reported values of urea (L.J. Bellamy et al 1975).

5.3.6 HOMO AND LUMO ANALYSIS

HOMOs and LUMOs are called as Frontier Molecular Orbitals; the energy difference between the molecular orbitals gives the band gap energy of the title compound. HOMO, LUMO and frontier orbital gap helps to illustrate the chemical reactivity and kinetic stability of the molecules. Atomic orbital HOMO-LUMO composition of the frontier molecular orbital of the title compound is represented in Fig. 5.5. The HOMO is the orbital that primarily acts as an electron donor and therefore the LUMO is that the orbital that primarily acts as an electron acceptor. HOMO and LUMO energies and orbital energy gap were computed at B3LYP/6- 311++G (d, p) method using the basic set. The calculated energy values for title compound in the gas phase are; E_{HOMO} = - 5.2312 eV, E_{LUMO} = - 2.1365 eV. HOMO–LUMO and Energy gap = 3.0947eV Table 5.5 displays the global reactivity descriptors of the LGY5N compound, which are determined using HOMO – LUMO values.

Table 5.4: The electric dipole moment μ, the average polarizability α_{tot} and first order hyperpolarizability β_{tot} for LGY5N molecule

Dipole Moment in Debye	
μ_x	2.18508
μ_y	-1.8269
μ_z	1.02354
μ_{tot}	1.17547
Polarizability in esu	
α_{xx}	24.1724
α_{xy}	0.36466
α_{yy}	21.7728
α_{xz}	3.1717
α_{yz}	3.9337
α_{zz}	11.6357
α_o	19.19363×10^{-24}
Hyperpolarizability in esu	
β_{xxx}	6.25545
β_{xxy}	2.02737
β_{xyy}	-1.51671
β_{yyy}	0.30574
β_{xxz}	-1.97626
β_{xyz}	4.39203
β_{yyz}	0.21664
β_{yzz}	-8.29667
β_{zzz}	-3.22623
β_{tot}	7.9010×10^{-30}

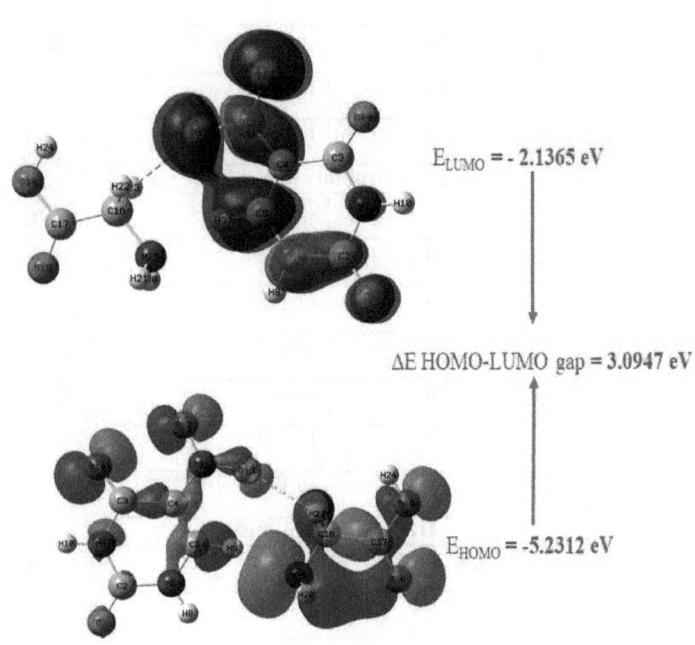

Fig. 5.5 HOMO – LUMO plot of LGY5N at B3LYP/6-311++G (d, p)

Table: 5.5 Calculated electronic and energies of LGY5N using B3LYP/6-311++G (d,p) level

Electronic parameter	Calculated values
E_{HOMO} (eV)	-5.2312
E_{LUMO} (eV)	-2.1365
$\Delta E_{HOMO} - E_{LUMO}$ (eV)	3.0947
Ionization Potential (IP)	5.2312
Electron Affinity (EA)	2.1365
Chemical Hardness (η)	1.5473
Softness (s)	0.3231
Electronegativity (χ)	3.6838
Chemical Potential (μ)	-3.6838
Electrophilicity index (ω)	4.3851

5.3.7 THERMODYNAMICAL PROPERTIES

The analysis of thermodynamic properties of a compound are extremely important in solid state science and are considered as an important trait in designing functional materials to be used under high temperature and high-pressure conditions. One of the key parameters of thermodynamics is the partition function which interlinks thermodynamics, spectroscopy and quantum theory. The standard statistical thermodynamic functions such as standard heat capacity (C_P), standard entropy (S), and standard enthalpy changes (ΔH) for the title compound were obtained from the theoretical harmonic frequencies on the basis of vibrational analysis at B3LYP/6–311++G (d, p) level using Thermo.pl software and listed in the table 5.6. From the observations, all the values of Cp, S and H increases with the increase of temperature from 100 K to 1000 K, which is accredited to the enhancement of the molecular vibration intensities. The temperature increases because at a constant pressure, the values of Cp, S and dH are equal to the quantity of temperature (F. Bopp et al 1967). The entropy of a system increases with increasing temperature since disorder increases with thermal agitation. The behaviour of materials under different thermodynamical constraints can be explained in terms of the specific heat of a solid. It also defines how efficiently the material stores heat. It is seen that at a lower temperature range, the heat capacity of title compound increases rapidly with increase in temperature Fig.5.8. Using quadratic formulas, the thermodynamic properties correlations are studied and temperatures are fitted. The corresponding fitting factors (R^2) were found to be 0.9993, 0.9901 and 0.9868 for entropy, heat capacity and enthalpy respectively.

$$S^0m = 0.0003x^2 + 0.114x - 6.958T \times 10^{-4}T^2 \ (R^2 = 0.9993)$$

$$CP^0m = 0.6579x^2 + 65.132x + 25\ T \times 10^{-4}T^2 \ (R^2 = 0.9901)$$

$$\Delta H^0m = 0.0003x^2 + 1.1598x + 260.75 \times 10^{-4}T^2 (R^2 = 0:9868)$$

Thermodynamic energies according to the relationships of thermodynamic functions can be calculated and approximate directions of chemical reactions in accordance with the second law of thermodynamics in thermochemical field can be evaluated (K. Bhavani et al 2015 and S. Muthu et al 2013).

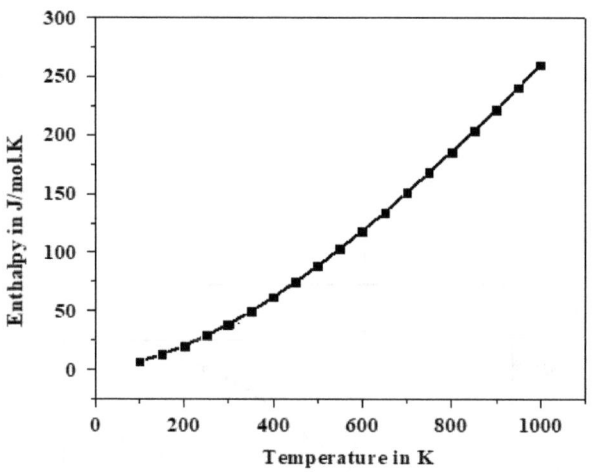

Fig. 5.6 Variation of Enthalpy with temperature

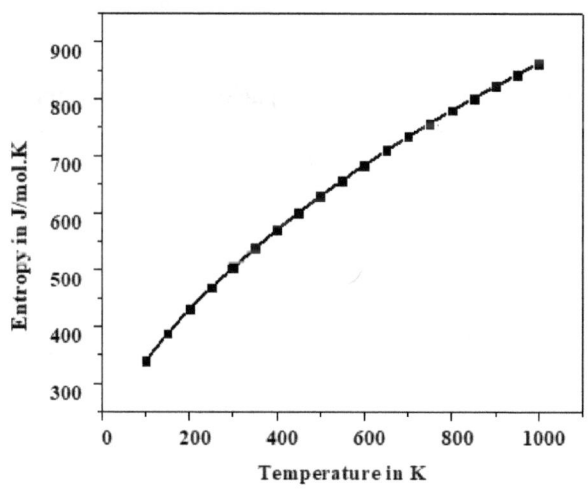

Fig. 5.7 Variation of Entropy with temperature

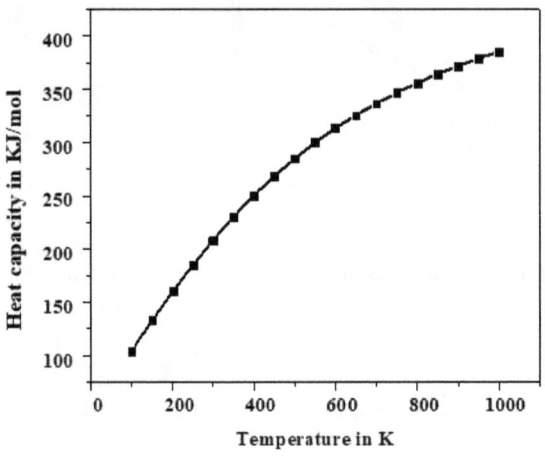

Fig. 5.8 Variation of heat capacity with temperature

Table 5.6: Thermodynamic properties at different temperatures by B3LYP level for LGY5N

T (K)	S (J/mol.K)	Cp (J/mol.K)	ΔH (kJ/mol)
200	498.112	186.251	26.214
250	512.214	212.365	39.526
300	574.253	284.271	47.062
350	624.021	312.254	69.251
400	672.236	522.895	86.712
450	718.247	378.236	98.209
500	771.895	406.253	118.414
550	801.245	426.254	136.239
600	825.147	444.231	156.144
650	878.125	456.695	182.166
700	912.354	471.254	202.122
750	929.256	484.733	235.124
800	974.214	501.321	266.845
850	998.286	515.524	284.179
900	1114.254	529.256	320.128
950	1042.274	538.365	325.858
1000	1074.264	552.321	344.121

5.3.8 OPTICAL STUDIES

The experimental UV-Visible spectrum of a LGY5N molecule was recorded by using methanol as solvent and it is compared with simulated spectrum calculated using TD-DFT/B3LYP method with 6-311++G (d, p) basis set. The maximum absorption (λ max) of our title molecule is estimated and also computed the band gap energy. The maximum peak λ max at 451nm showing a good match with measured experimental data of 412 nm. The band gap energy was estimated using the formula, $E = hc/\lambda$. Here h and c are constant; λ is the cut off wavelength. The UV-Visible absorption spectrum of LGY5N is shown in Fig.5.9. The optical band gab energy value is found to be 2.85 eV is shown in Fig.5.10. The spectrum reveals that the transmittance of 1 mm thickness LGY5N crystal is ~ 93% in entire range of spectrum, further it is observed that the transmission of crystal sharply falls to minimum value in lower wavelength region. The theoretical band gap energy value has been calculated by the direct band gap method which is 2.85 eV. It shows that the theoretical values are in good agreement with experimental data. Figs. 5.12, 5.13 and 5.14, shows the Extinction coefficient, refractive index and reflectance are optical parameters that are used to determine the material effectiveness in the manufacture of optoelectronic device.

5.3.9 MULLIKEN POPULATION ANALYSIS

The Mulliken charge is directly related to the vibrational properties of the molecule and quantifies how the electronic structure changes under atomic displacement; it is therefore related directly to the chemical bonds present in the molecule. It affects dipole moment, polarizability, electronic structure and more properties of molecular systems. The Mulliken population analysis in LGY5N molecule was calculated using B3LYP level with 6-311++G (d, p) basis set. The Mulliken charge distribution structure of LGY5N is shown in Fig.5.15. The charge distribution of LGY5N shows that the carbon atoms attached with hydrogen atoms are positively charged. Whereas the other carbon atoms are negative. All hydrogen atoms are positively charged. Moreover, Mulliken atomic charges also show that the H atoms attached to N or O atoms have bigger positive atomic charges than the other hydrogen atoms. This is due to the presence of electronegative nitrogen and oxygen atoms. Carbon is the topmost positive charge possessed atom (~1.6882) among the other positive charge possessed atoms like H1, H2 and H8 is the most negative charge possessed atom (~1.982) among the other negative charge possessed atoms like N, O1, O6 and C4 in the molecule. Due to the electron drop out character, Carbon atom keeps the extreme positive charge.

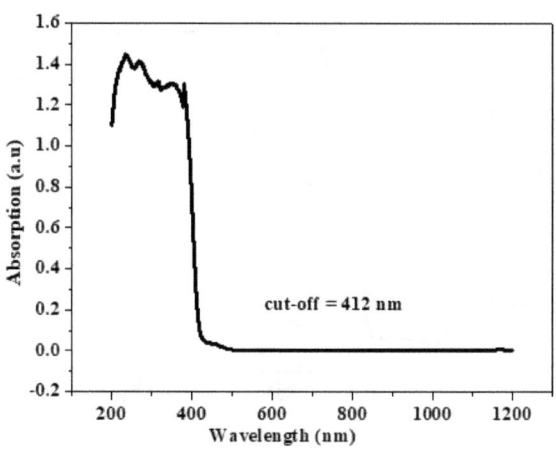

Fig. 5.9 UV- Vis absorption spectrum LGY5N of crystal

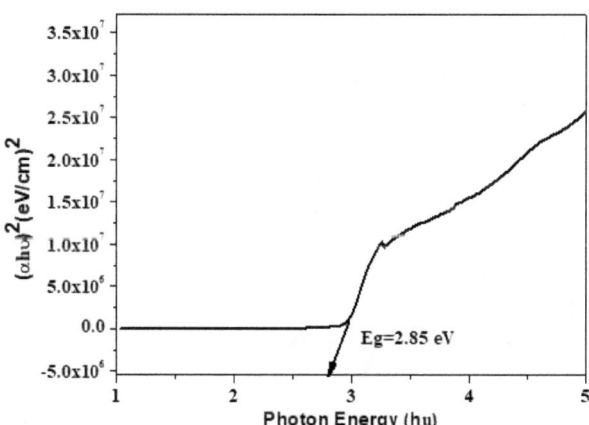

Fig. 5.10 UV- Vis bandgap plot of LGY5N crystal

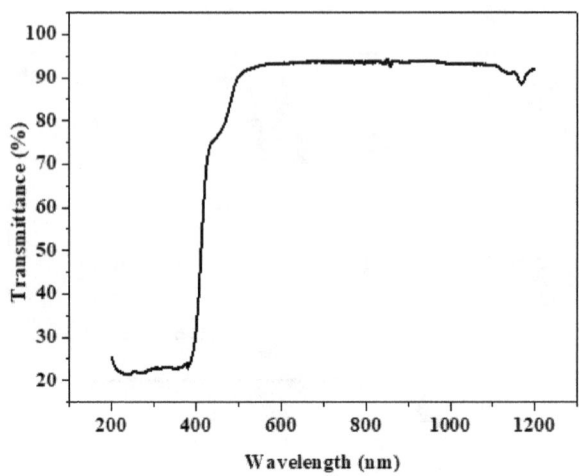

Fig. 5. 11 UV-Vis transmission spectrum of LGY5N crystal

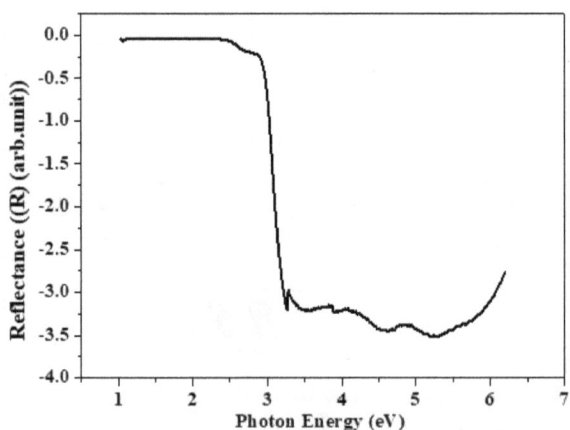

Fig. 5. 12 UV- Vis reflective index of LGY5N crystal

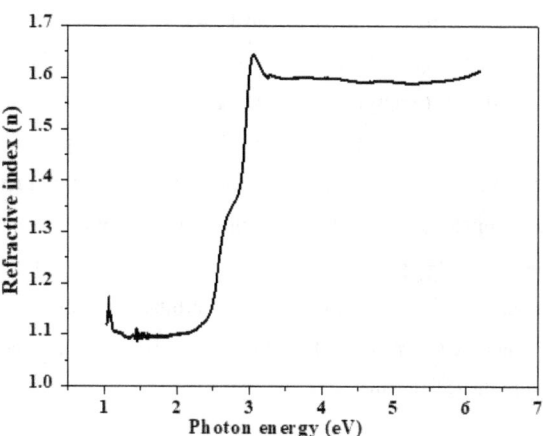

Fig. 5. 13 UV- Vis refractive index of LGY5N crystal

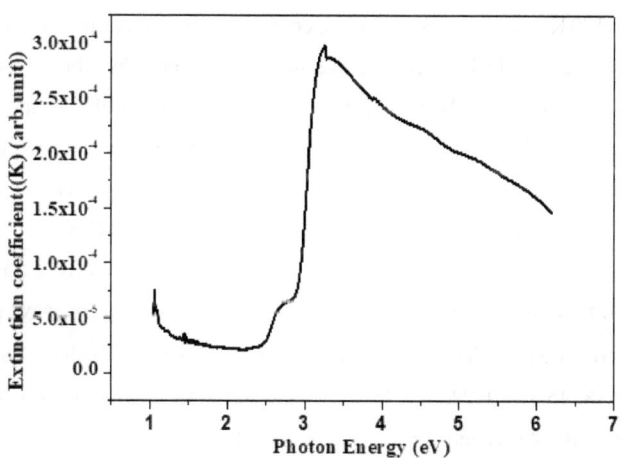

Fig. 5. 14 UV-Vis extinction coefficient of LGY5N crystal

5.3.10 NATURAL BONDING ORBITAL (NBO) ANALYSI

The NBO 3.1 programme which was run in the Gaussian 09 software at the DFT (B3LYP) / 6-311++ G (d, p) stage, it used to investigate the interactions that occur between the leiws (filled) and non-lewis (empty) orbital, which quantity the delocalization or hyper conjugation interactions (Reed A.E et. al. 1988). The LGY5N compound was subjected to NBO analysis to assess the possible intramolecular and intermolecular interactions between the lewis and non-Lewis orbitals. In table 5.8, the stabilisation energy $E^{(2)}$, which was calculated continuously by the 2nd order perturbation principle (Glendening E.D et.al. 2003), was associated with the relationship connecting the filled orbital i and the empty orbital j. The higher $E^{(2)}$ value, greater the interaction between the donor (i) and the acceptor (j). the formula can be used to measure the $(E^{(2)})$ value.

$$E^{(2)} = q_i \frac{(i,)^2}{\Sigma_j - \Sigma_i}$$

Where, q_i is the donor orbital's occupied stage. Σ_i Σ_j are the energies of the donating and accepting orbital's and F(i,j) is the NBO Fock matrix compounds (Muhammad Khalid et. al. 2020). Hydrogen bonding interactions, which have a wide range of stabilisation energies ranging from 2 to 170 KJ/mol, have essential properties in system structuring. The stabilisation force in the system is caused by hyperconjugative interactions framed by orbital overlap among occupied and unoccupied neighbouring orbitals. table 5.8, shows the focused propable interaction of bonding orbital (donor) and antibonding orbital (acceptor) with enormous second order perturbation energy E(2). DFT (B3LYP) / 6-311++ G (d, p) was used to calculate the NBO. For LGY5N, the strong intramolecular interaction are π (C4-C5) → π^* (C3-O14), π(N11-O12) → π^*(C4-C5), π (O7) → σ^*(N1-C2), π (O7) → σ^*(C2-N6), π (O12) → π^* (N11-O12), π (O13) → σ^*(O13-H23), π (O14) → σ^* (N1-C3), π(O14) → π^*(N11-O12), π^* (N11-O12) → π^*(C4-C5), σ^*(O13-H23) → σ^*(H23) with higher $E^{(2)}$ values, 26.72, 27.49, 28.99, 28.62, 12.48, 15.36, 20.22, 172.46, 10.90 and 82.54 Kcal/mol correspondingly. The bonding between π^*(C3-O14) and the acceptor of π^*(C4-C5) has a strong interaction with a stabilisation energy of 193.35 Kcal/mol. These transitions occur within molecules with a high polarisation, implying that LGY5N has more NLO activity (Glendening E.D et. al. 2011). Table 5.8, lists the rest of the documents. These results shows that the LGY5N compound has a high level of stabilisation capacity.

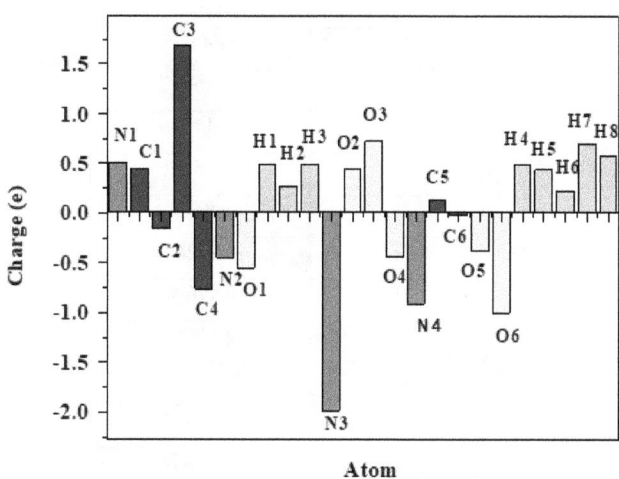

Fig. 5.15 Mulliken atomic charges of LGY5N single crystal

Table 5.7: Mulliken atomic charges of LGY5N single crystal

Atoms	Mulliken atomic charge
N1	0.50603
C1	0.44605
C2	-0.15782
C3	1.68825
C4	-0.76027
N2	-0.44336
O1	-0.55166
H1	0.49101
H2	0.27265
H3	0.49634
N3	-1.98293
O2	0.44229
O3	0.72804
O4	-0.43414
N4	-0.9073
C5	0.13778
C6	-0.02312
O5	-0.38215
O6	-0.99631
H4	0.48584
H5	0.44197
H6	0.22789
H7	0.70096
H8	0.58603

5.3.11 MICROHARDNESS TEST

The microhardness gives the knowledge of the strength, yield stress and deformation features of the material (Kishan Rao et al 1983 and Subhadra, K. G al 2000). The microhardness measurements on the well-polished smooth surface of the crystal cut along the plane (0 1 0) was implemented. Measurements were taken for different loads such as 5, 25, 50 and 100 g in room temperature with the indentation time 10s. The Vicker's hardness number (Hv) was related to the applied load (P) and the diagonal length of the indentation (d) as, (Vengatesan et al 1986)

$$H_v = 1.8544 P/d^2 \text{ kg/mm}^2$$

The hardness number of different loads was estimated from the above Equation and a graph is plotted as illustrated in Fig.5.16. It is noticed that there is an increase in H_v with the increase in load and this symbolizes the Reverse Indentation Size Effect (RISE). The plot of log P versus log d given in Fig. 5.17. gives a straight line which is in good agreement with the Meyer's law, $P = k_1 d^n$ (E. Meyer et al 1908), where k_1 is the material's constant and n is the Mayer's index called work hardening efficient. The value of n is determined from the slope of the straight line. According to the Onitsch (E.M. Onitsch et al 1950) classification, the value of (n =5.29) suggests that the LGY5N crystal belongs to a soft material category.

5.3.12 THERMAL ANALYSIS

The thermogravimetric investigation of the LGY5N was carried out in the temperature range of 50–600 °C. The DTA/TGA curve of the LGY5N was depicted in Fig.5.18. In DTA the peaks at 205°C and 457°C represented the decomposition point and melting point of the compound. The sharp endothermic peak showed a good degree of crystallinity of the present compound. TGA curve of the present compound showed that the material was stable up to 148°C and there is no phase transition up to 148°C, and it is also affirmed the absence of water molecules during the crystallization process. Weight loss begins at 148°C and proceeded up to full evaporation of the materials. The first stage of weight loss happened between the temperatures 177°C with 35%, which was due to the discharge of volatile substances. This decomposition process is revealed by a strong endothermic peak on the DTA curve, with the maximum being at 211°C.

Table 5.8: Second order perturbation theory analysis of Fock Matrix in NBO for LGY5N with 6-311++G (d, p) basis set

Donor (i)	Type	Acceptor (i)	Type	E(2)^a (k cal/mol)	E(j)-E(i)^b (a.u)	F(I,j)^c (a.u)
C3-C4	σ	N11-O13	σ*	18.03	1.12	0.127
C4-C5	π	C3-O14	π*	26.72	0.25	0.079
N6-H8	π	C4-C5	π*	10.48	1.29	0.104
N11-O12	π	C4-C5	π*	27.49	0.33	0.091
N1	n1	C2-O7	π*	65.01	0.27	0.119
N1	n1	C3-O14	π*	89.22	0.25	0.134
N6	n1	C2-O7	π*	61.73	0.24	0.111
N6	n1	C4-C5	π*	43.06	0.25	0.097
O7	n1	C2	σ*	19.19	1.56	0.155
O7	π	N1-C2	σ*	28.99	0.63	0.123
O7	π	C2-N6	σ*	28.62	0.64	0.123
O12	π	N11-O12	π*	12.48	0.08	0.040
O13	π	N11-O12	π*	15.36	0.30	0.082
O13	π	O13-H23	σ*	51.34	1.28	0.235
O14	n1	C3	σ*	13.00	1.59	0.129
O14	π	N1-C3	σ*	20.22	0.83	0.127
O14	π	N11-O12	π*	172.46	0.09	0.138
C3-O14	π*	C4-C5	π*	193.35	0.03	0.095
N11-O12	π*	C4-C5	π*	10.90	0.19	0.045
O13-H23	σ*	H23	σ*	82.54	0.05	0.136

From the thermal study results it was concluded that the crystal was thermally stable up to 177°C and it could be utilized for any suitable applications up to this temperature.

5.3.13 DIELECTRIC STUDIES

The analysis of dielectric response of crystal provides the insight of materials liability to find application in microelectronics and optoelectronics devices. The major parameters that are vital to investigate are dielectric constant and dielectric loss. The dielectric constant in a bulk crystal material is contributed by the electronic, ionic dipolar and space charge polarization (B.S. Mitchell et al 2004) which usually show highly active response (high dielectric constant) at lower frequencies and minimized effect (low dielectric constant) at higher frequencies (M. Anis et al 2017 and M. Kumar et al 2017). The dielectric constant (Fig.5.19) becomes least responsive beyond 1 KHz frequency owing to which it is observed that the LGY5N crystal offers significant decreases in dielectric constant with increase in frequency at all applied temperatures.

The contribution of low dielectric constant comes with the benefit that it impedes power consumption in medium and reduces the σc delay which is the most essential factor when designing optoelectronics and microelectronic devices (R. Subhashini et al 2016 and B.D. Hatton et al 2006). The presence of macro/micro cracks, random crystal orientation, voids, impurities, inclusions in crystal encourages the nucleation of dielectric loss (S.P. Ramteke et al 2018). The dielectric loss of LGY5N crystal has been investigated with reference to the frequency as shown in Fig.5.20. It is observed that the dielectric loss decreases with increases in frequency at respective temperature. This typical decreasing nature of dielectric loss indicates that the LGY5N crystal possesses good optical quality and minimum electrically active defects (R.N. Shaikh et al 2016). The designing of photonics, NLO, electro-optic modulators, field detectors and THz wave generators demand material with low dielectric properties (A. Bhaskaran et al 2010 and S. Ramteke et al 2018) which suggest LGY5N crystal as promising candidate for device fabrication.

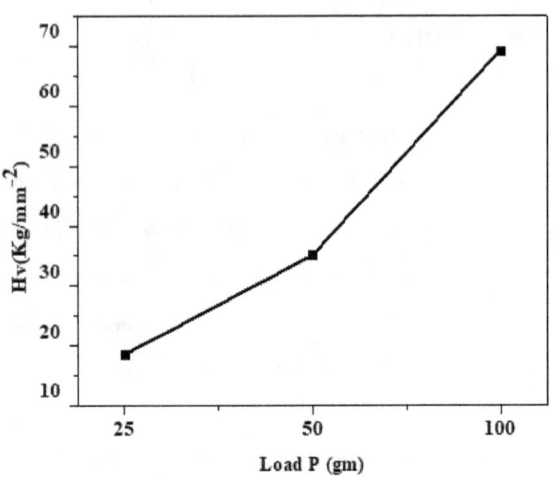

Fig. 5.16 Vickers hardness number Vs applied load of LGY5N

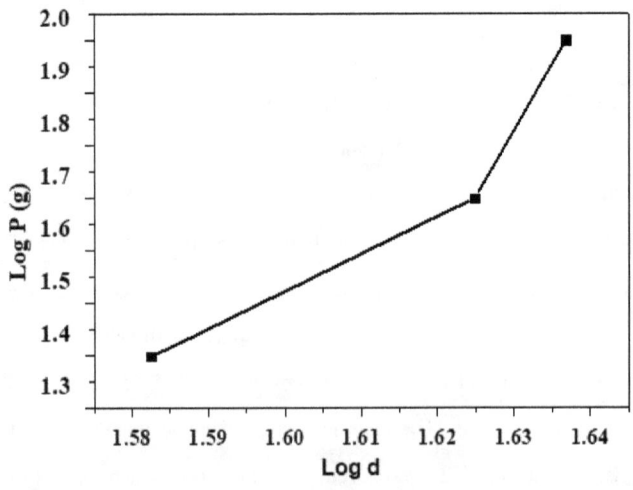

Fig. 5.17 log p Vs log d plot of LGY5N

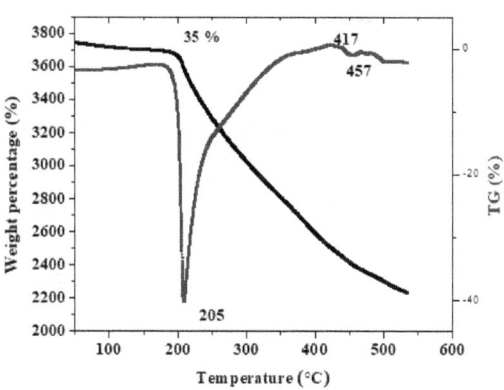

Fig. 5.18 TG-DTA plot of LGY5N crystal

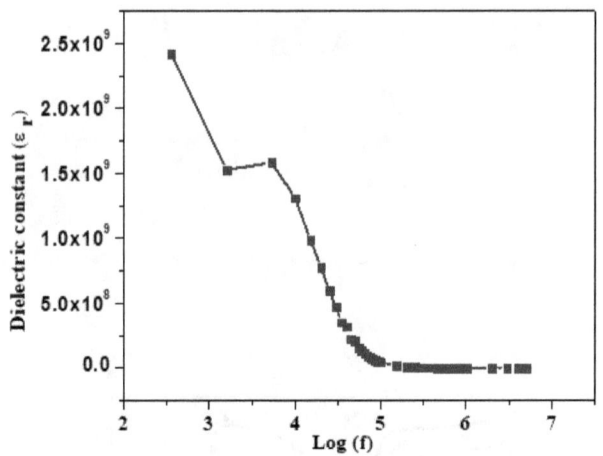

Fig. 5.19 Variation of dielectric constant of LGY5N

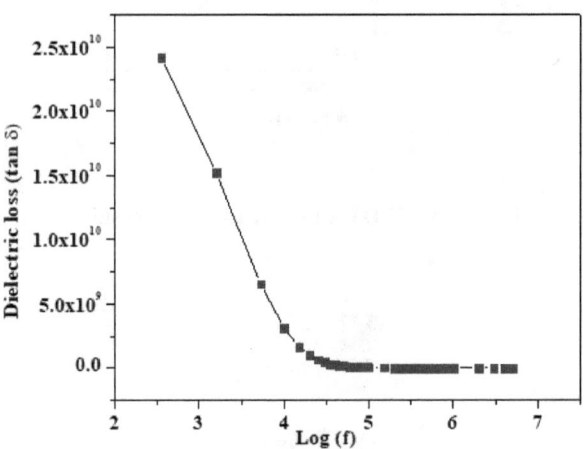

Fig. 5.20 Variation of dielectric loss of LGY5N

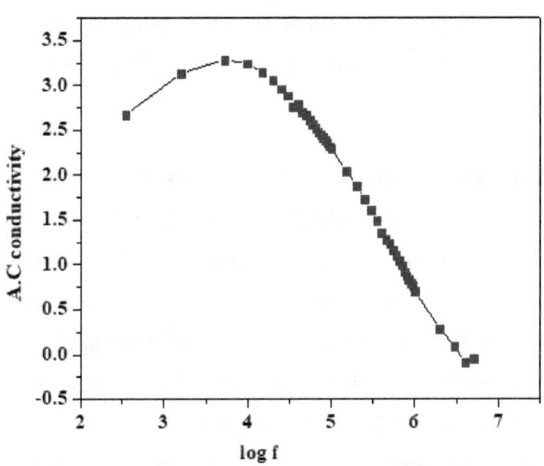

Fig. 5.21 Frequency dependence of AC Conductivity

5.3.14 PHOTOLUMINESCENCE ANALYSIS

PL emission spectra can be used to identify the surface, interface, impurity levels, disorder and interface roughness while the intensity provides the quality of crystal surfaces and interfaces. To explore the potential applications as effulgent materials, solid-state fluorescent properties of the material was studied. The photoluminescence spectrum of crystalline powder recorded at room temperature with excitation at 236 nm is shown in Fig.5.22. The material shows the emission band entered at 456 nm and this luminescence probably may emerge from the protonated 5- Nitrouracilate and hence deputized to an intraligand charge-transfer (ILCT) transition. The high intensity peak observed at 485 nm indicates that the crystal has a blue emission and hence the LGY5N crystal can be a suitable material for LED's applications.

5.3.15 MOLECULAR ELECTROSTATIC POTENTIAL

The Molecular Electrostatic Potential (MEP) at a given point p (x,y,z) in the vicinity of a molecule is determined by the force acting on a positive test charge (a proton) located at a point through the electrical charge cloud generated through the molecules electrons and nuclei (Murray J.S et. al. 1991 and Luque F.J et al 2000). The MEP is typically visualized through mapping its values onto the surface reflecting the molecules boundaries thereby providing a visual method to understand the relative polarity of a molecule.

Electrostatic potential correlates with dipole moment, electronegativity and partial charges and also illustrate information about the charge distribution of a molecule, electrostatic potential properties of the nucleus and nature of electrostatic potential energy. MEP is correlated to the electronic density and is a very useful descriptor in finding sites for electrophilic and nucleophilic reactions as well as hydrogen bonding interactions (Politzer P et al 1990). MEP was calculated at the B3LYP/6–311++G (d, p) to attain optimized geometry. The 3D plots of the MEP of the title compound are illustrated in Fig. 5.23.

The different values of the electrostatic potential at the surface of the molecule are represented by different colours. Increases in potential are identified by colour in the order of red b orange b yellow b blue. The colour code of these maps ranges from −9.388 eV and +9.388 eV, where blue indicates the strongest attraction and red indicates repulsion. The majority of light green region MEP surface depicts a potential halfway between two extremes red and dark blue colour. As can be seen from the MEP of the title compound, while the regions with negative potentials are over the electronegative atoms (carbon, nitrogen and oxygen) and the regions with positive potentials are over the hydrogen atoms.

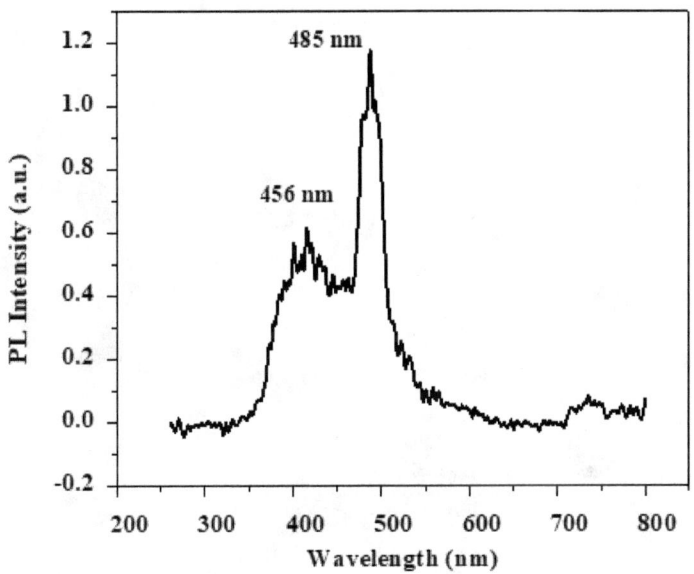

Fig. 5.22. Fluorescence emission spectra of LGY5N crystal

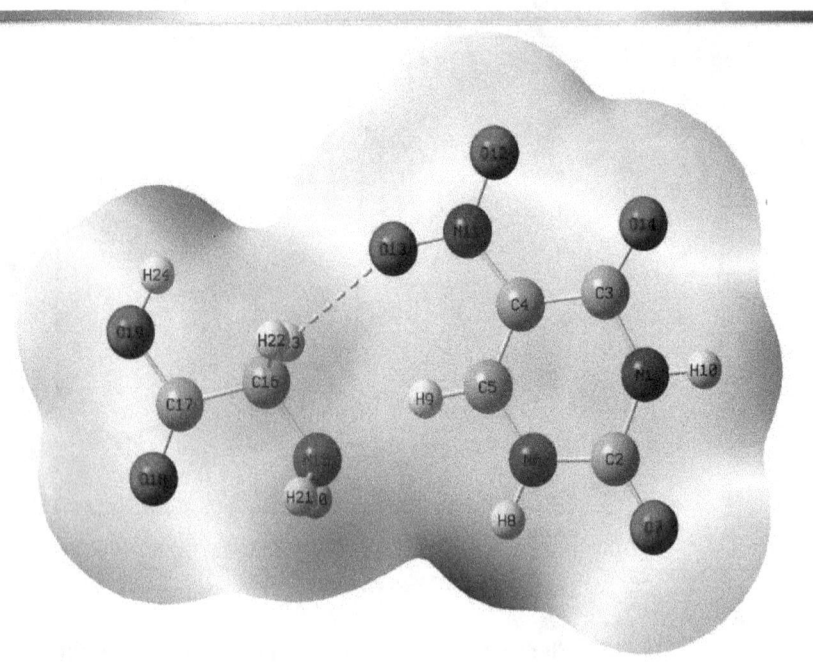

Fig. 5.23. Molecular electrostatic potential map (MEP) of LGY5N crystal

5.3.16 SECOND HARMONIC GENERATION (SHG)

Organic materials are expected to have relatively strong nonlinear optical properties due to delocalized electrons at π-π* orbitals. This expectation explains extensive search for better NLO materials among organic materials. The title compound being an organic material shows good nonlinear optical properties obtained from the calculation using B3LYP/6-311++G (d, p) method. The output energies from the grown sample and reference KDP are found to be 20 mV and 35 mV respectively. The results show that SHG efficiency of the grown crystals was 1.7 times that of the standard Potassium dihydrogen phosphate (KDP) crystal.

5.4 CONCLUSION

Researchers used both experimental and theoretical spectroscopic investigations to obtain the results from employing FT-IR, a method that was derived from DFT. Overall, the experiment and theoretical modes of vibration were in agreement. DFT/B3LYP calculations, employing a 6-311++G (d, p) basis set, gave vibrational frequencies and infrared intensities for the optimized molecular geometry. The energy gap of the molecule was calculated using HOMO and LUMO analyses, and it was found to be 3.0947 eV. The very strong band in the absorption spectrum was found at 431nm in UV spectrum. The energy of the band gap was estimated by 2.84 eV. The energy difference between the HOMO-LUMO energy gap is generally the lowest energy electronic excitation that is possible in a molecule. The energy gap of the HOMO-LUMO can tell us about what wavelengths the compound can absorb. Smaller HOMO-LUMO energy gaps correspond to better stability. The term band gap in UV refers to the energy difference between the top of the valence band to the bottom of the conduction band. The electrons are able to jump from one band to another. The terms band gap and energy gap cannot be used interchangeably. The variance of reflectance (R) with the wavelength and the difference of refractive index (n) and the extinction coefficient (k) with photon energy of the LGY5N crystal with respect to wavelength were also investigated. Meyer's test helps to identify the LGY5N crystal belongs to a soft material category. The endothermic peak was detected in the process of breakdown with the highest occurring at 211°C. The intermolecular charge transfer of the sample is explained by NBO and mulliken charge analysis. At room temperature, the dielectric constant was explained as a function of frequency using the dielectric loss graph. The efficiency of SHG in grown crystal is 1.7 times that of reference KDP.

CHAPTER VI

SUMMARY AND SUGGESTIONS FOR FUTURE WORK

6.1 INTRODUCTION

In the past few decades, interests have been made to grow good quality nonlinear optical crystals to meet several requirements, such as large phase matching angle, large nonlinear optical coefficient and a wide optical window around the visible region, good mechanical and chemical stability and high laser damage threshold. Organic NLO materials offer several advantages, such as large nonlinear figure of merit, high optical damage threshold, fast response time and their structural flexibility facilitates easy control of physical properties over very wide range. But their moderate environmental stability, low mechanical strength and a limited temperature range of operation limits their application.

The present comprises of growth and characterization of work LAHSQ, LGHSQ, LA5N AND LGY5N single crystals.

6.2 SUMMARY

Chapter I gave a brief introduction about crystal growth methods, experimental techniques, computational methods and nonlinear optical materials. In computational method the DFT was briefly explained. This gave a detailed idea about bonding and vibrational frequencies. The different analytical, spectroscopic techniques and experimental techniques were used for the investigation of the four materials. For the characterization of these materials, we have used Single X- ray diffraction, FT-IR, UV spectral studies, Micro hardness studies, Thermal studies, Dielectric studies and SHG efficiency. DFT calculations were performed using Gaussian 09 software with 6-311++G (d, p) basis set.

Chapter II describes synthesis and growth of single crystals of LAHSQ was done in aqueous solution using slow evaporation technique at room temperature. From the single crystal X-ray diffraction study, the lattice parameters of LAHSQ crystal were measured as, a =5.1301Å, b =8.3224Å and c =14.9005Å, V=621 (Å^3) $\alpha \neq \beta \neq \gamma \neq 90°$ and cell volume (V) = 621Å3 which reveals LAHSQ crystal has triclinic crystal structure. The crystalline nature of the grown crystal was identified by single crystal XRD analysis. The various modes of vibrations were identified for LAHSQ crystal by FT-IR studies. A computation of LAHSQ molecule calculated gives the optimized structure. Bond Length and Bond Angle are calculated theoretically by DFT method. The optical nature of the grown single crystal was evaluated from the absorption spectrum. Theoretical and experimental IR spectroscopic analysis was performed to identify the functional groups in LAHSQ molecule. HOMO-LUMO analysis revealed the molecular energy gap. Other than this first hyperpolarizability, thermodynamic parameters and atomic charges were calculated using 6-311++G (d, p) basis set. From UV-Vis studies, it is witnessed that LAHSQ has minimum absorption upto 1200 nm, which is the key requirement for materials having NLO applications. UV-Vis studies of LAHSQ reveal a lower cut-off wavelength of 364 nm, a basic need for NLO materials. Using Tauc's map the optical bandgap value was 3.20 eV. The absorption coefficient (α), reflectance (R), Extinction coefficient (K) and the refractive index (n) were determined. The mechanical study proves the work hardening coefficient of LAHSQ Vickers microhardness study led to the evaluation of the hardness number (Hv) approach and proved LAHSQ as belonging to the soft material category. The TG/DTA analyses confirmed that the material is stable up to 398°C. From the dielectric studies it can be concluded that the variation of dielectric constant is having an inverse relation at low frequencies and gets saturated at high frequencies. The PL spectrum shows blue emission in the crystal. MEP for the LAHSQ molecule is calculated by B3LYP level with the 6–311++G (d, p) method. The innteractions that occur in the molecule were also discussed in the mulliken charge estimations and NBO investigation. The NLO test confirmed the superiority of conversion efficiency of LAHSQ single crystal over KDP, which is about 5.5 times that of reference material.

Chapter III describes the synthesis and growth of good optical quality L- Glycinium Hydrogen Squarate single crystal have been grown successfully by SEST. Single crystal XRD studies were carried out for the grown crystal. The single crystal analysis shows that the crystal belongs to Monoclinic crystal system with P2$_1$ space group. (DFT) computations of LAO molecule calculated by DFT (BLYP) level with 6-311++G (d, p) basis set gives the optimized structure. Experimentally obtained bond lengths and bond angles are compared with theoretically calculated one. The important functional groups of LGHSQ crystal were identified and confirmed the composition of grown crystal by FT-IR studies. Theoretical and experimental IR spectroscopic analysis was carried out and the presence of functional groups in LGHSQ molecule was qualitatively analyzed. HOMO-LUMO analysis reveals the molecular energy gap. From, the thermodynamic functional analysis can be used to calculate the values of thermodynamical properties, thermodynamic quantities in particular heat capacity, entropy and enthalpy changes for synthesized LGHSQ single crystal have been measured by the vibrational analysis at B3LYP level with the 6–311++G (d, p) basis set. The UV-Visible spectrum of the LGHSQ crystal has shown that they are highly transparent in the visible region and the cut-off wavelength is 286 nm. The calculated direct optical bandwidth value was 4.62 eV. In order to analyze the optical properties, the absorption coefficient (α), reflectance (R), extinction coefficient (K) and refractive index (n_0) have been calculated. The natural population analysis (NPA) of the L-glycinium hydrogen squarate (LGHSQ) molecule was found by Mulliken Population Analysis (MPA) using B3LYP 6–311++G (d, p) basis set. The vickers microhardness study led to the evaluation of the hardness number (Hv) approach and proved LGHSQ as belonging to the soft material category. Thermal stability of LGHSQ is found. The dielectric constant, dielectric loss and AC conductivity was studied as a function of frequency at room temperatures. The PL spectrum shows blue emission in the crystal. MEP for the LAHSQ molecule is calculated by B3LYP level with the 6–311++G (d, p) method. The intermolecular charge transfer is registered using NBO analysis. Investigations of dielectric constant, dielectric loss, and AC conductivity as a function of frequency were conducted at room temperature. The second harmonic generation of the grown crystal was measured and compared with KDP. Thus, with many attaching linear and nonlinear optical properties, it is concluded that the synthesized L- Glycinium Hydrogen Squarate is a suitable candidate for optoelectronic application.

Chapter IV Employing SEST optically good quality single crystals of LA5N were grown. The single crystal XRD study confirms the unit cell parameters of LA5N are a =8.8632 Å; b =9.9658 Å; c =15.8218 Å. α = γ = 90° and β= 96.16. From SXRD analysis the grown LA5N single crystal is determined to be monoclinic crystalline nature. Optimization of LA5N single crystal is performed by using DFT approach. Theoretically calculated vibrational frequencies are compared with experimentally obtained FT-IR frequencies. Spectral assignments are carried out for various vibrational frequencies. The magnitude of molecular first order hyperpolarizability for this molecule is in suitable range to exhibit NLO behavior. Molecular energy gap of LA5N was found as 2.6985 au by HOMO-LUMO analysis. The standard thermodynamic functions can be used as reference thermodynamic values to calculate the changes of entropies (ΔS), changes of enthalpies (ΔH) and changes of Heat capacity (ΔC) of the reaction. The UV-Visible spectrum of the LGY5N crystal showed that they are highly transparent in entire visible region, with the cut-off wavelength of 431 nm. The calculated optical direct band-gap value was estimated at 2.94 eV from Tauc's plot. Optical parameters such as absorption coefficient (α), reflectance (R), extinction coefficient (K) and refractive index (n) were calculated to analyse the optical properties. This optical analysis reveals the NLO behaviour of the material. Mulliken atomic charge calculation helps in understanding the chemical potential and ionization potential of the molecule. The intermolecular charge transfer is registered using NBO analysis. Mechanical strength, Thermal stability, photoluminescence analysis, dielectric studies, vickers microhardness studies and SHG studies were carried out and the conversion efficiency of the LA5N crystal is found to be 1 time that of KDP crystal. Low values of dielectric constant and loss at high frequency confirms the polarizing behavior of the candidate material.

Chapter V This chapter deals with the synthesis, growth and characterization of nonlinear optical single crystals of L-Glycinium 5-Nitrouracilate (LGY5N) Single crystal were grown from supersaturated solution by slow evaporation technique. The single crystal X-ray diffraction study confirms the structure of as grown LG5N crystal and the unit cell parameters are a=7.8254 Å, b=8.2541 Å, c= 14.528 Å, α = γ=90°, and β= 93.417°. The Structure of LGY5N single crystal was monoclinic and the space group was $P2_1$. Optimization of LGY5N single crystal is performed by using DFT approach. FT-IR analysis confirmed the grown crystal. FT-IR spectrum was compared with the quantum computational methods using 6-311++G (d, p) basis set. Theoretical values are in good agreement with the experimental values. First order hyperpolarizability of LGY5N is calculated by two different basis sets and found useful in molecular designing. Band gap energy is found by HOMO – LUMO is 3.0947 eV by DFT basis set. From, the thermodynamic functional analysis can be used to calculate the values of thermodynamical properties, thermodynamic quantities in particular heat capacity, entropy and enthalpy changes for synthesized LGY5N single crystal have been measured by the vibrational analysis at B3LYP level with the 6–311++G (d, p) basis set. The UV-Visible spectrum of the LGY5N crystal showed that it is highly transparent in entire visible region, with the cut-off wavelength of 412 nm. The calculated optical direct band-gap value was estimated at 2.85 eV from Tauc's plot. The Mulliken population analysis in LGY5N molecule was calculated using B3LYP level with 6-311++G (d, p) basis set. The intermolecular charge transfer is confirmed by NBO analysis. The microhardness studies indicate that LGY5N belongs to soft category of materials. The thermal behaviour of the material analyzed by TG/DTA suggests that LGY5N is stable up to 205°C. The dielectric constant is found to be high at lower frequencies and then it decreases with increase in frequency. Further, the dielectric loss also decreases with increase in frequency. The low value of dielectric loss of the crystal is an indicator to the fact that the tested specimen is having less number of defects. AC conductivity were studied at room temperatures as a function of frequency. The LGY5N crystal PL spectrum shows UV light, blue light emissions that indicate their high structural and optical quality. The crystal NLO activity measurement indicates that the LGY5N has a SHG efficiency of 1.7 times that of KDP. So, we can conclude that the synthesized compound is suitable for optoelectronic applications.

6.3 SUGGESTIONS FOR FUTURE WORK

Additional evaluations are possible based on the outcomes of the current examinations, which are sketched below.

- In future, the alternative sophisticated methods like Sankaranarayanan-Ramasamy (SR) technique, Czochralski technique and Bridgeman technique might probably be an attempt to develop large crystal of LAHSQ, LGHSQ, LA5N and LGY5N compounds for industrial usage.
- The title compounds will be included nucleation studies such as, super saturation, meta-stable zone, induction period, critical radius and interfacial energy for the enhanced growth conditions appropriate for developing the large crystal with advancement.
- The NLO character probably is stimulated in the doping of crystals with suitable metals to be investigated. Piezo-electric and electro-optic effectiveness should have the biggest wonder, whenever actualized on the doped crystals. The thermal and electrical conductivity along different planes will be tried. Endeavours to assess the birefringence, specific heat capacity and phase matching angle of developed crystals illuminating and it will be attempted to use title samples in NLO.
- Structure and imperfection of the crystals can be envisioned by utilizing Atomic Force Microscope (AFM) and Scanning Electron Microscope (SEM). The etching character of different crystallographic faces of the crystals can be inspected with various organic solvents, to make out the dislocation, lattice in-homogeneity and afterward to process crystallinity.
- Amino mixtures are a fascinating compound for NLO as well as a few biomedical applications. Finally, future attempts to build electro optical components LAHSQ, LGHSQ, LA5N and LGY5N compounds can be created.

REFERENCES

1. Aggarwal M.D, Stephens J, Batra A.K and Lal R.B, Journal of Optoelectronics and Advanced Materials. 5 (2003) 555-562.
2. Aiping F, Dongmei D and Zhengyu Zhou, Spectrochimica Acta Part A: Journal of Molecular and Biomolecular Spectroscopy 61 (2003) 245-253.
3. Altun A, Golcuk K and Kumru M, Journal of Molecular Structure. 637 (2003) 155-169.
4. Ambujam K, Preema C, Thomas, Aruna S, Prem Anand D and Sagayaraj P, Journal of Materials Manufacturing Processes. 22 (2007) 346-350.
5. Amudha M, Madhavan J and Praveen Kumar P, Journal of Optics. 46 (4) (2017) 382-390.
6. Anis M, Hussaini S.S, Shirsat M.D, Shaikh R.N, and Muley G.G, Journal of Materials Research Express. 3 (10) (2016) 106204-106209.
7. Anis M, Pandian M, Baig M, Ramasamy P and G Muley, Taylor & Francis 22 (2018) 409-114.
8. Anis M, Shkir M, Baig M.I, Ramteke S.P, Muley G.G, AlFaify S and Ghramh H.A, Journal of Molecular Structure 1170 (2018) 151-159.
9. Ardle M.C, Sherwood J.N and Damask A.C, Journal of Crystal Growth. 22 (1974) 193-200.
10. Arjunan S, Bhaskaran A, Kumar R. M, Mohan R and Jayavel R, Journal of Alloys and Compounds. 506 (2) (2010) 784-787.
11. Arulmani S and Senthil S, Journal of Materials Today Proceedings 5 (2020) article in press.
12. Arulmani S, Senthil S, Journal of Materials Today Proceedings 5 (2018) 8996-9003.
13. Arulmani S, Venkatesan A, Rajasaravanan M.E and Senthil S, Journal of Materials Today Proceedings 8 (2019) 136-141.
14. Badan J, Hierle R, Perigaud A and Zuss J (Eds.), American Chemical Symposium Series 233, Journal of American Chemical Society, Washington. DC, 1993.
15. Bahgat K and Ragheb A, Central European Journal of Chemistry 5 (1) (2007) 201-220.
16. Baig, M.I, Anis M and Muley G.G, Journal of Optical Materials 72 (2017) 1-7.
17. Balarew C, Duhlew R, Journal of Solid Sate Chemistry, 1 (1984), 55.
18. Becke A.D, Journal of Chemical Physics. 98 (1993) 5648-5652.

19. Bellamy L.J, The IR spectra of Complex Molecules., John Wiley and Sons, NY, 1975.
20. Bevan Ott and Boerio- Goates Journal of Applied Mechanics reviews 54 (2001) 664-675.
21. Bhavani K, Renuga S, Muthu S, and Sankara narayanan K, Spectrochimica Acta Part A: Journal of Molecular and Biomolecular Spectroscopy. 136 (2015) 1260-1268.
22. Bin Jiang and Zhilu Liu, Journal of Acta Crstallographica 65 (2009) 2232.
23. Boopathi K, Moorthy Babu S, Jagan R, Athimoolam S, Ramasamy, New Journal of Chemistry. 21 (2018) 1-34.
24. Bopp F, Zeitschrift Fur Journal of Physik. 200 (1967) 142-157
25. Buckley H. E, Crystal Growth John Wiley and Sons, New York. 1951.
26. Chermette H, Comput Bopp F, Zeitschrift firr Journal of Physics B.200 (1967) 117-132
27. Colthup N.B, L.H. Daly and S.E. Wiberley., Introduction to Infrared and Raman Spectroscopy, Academic Press, New York, 1990.
28. Colthup N.B. Daly L.H, Wiberley S.E, Academic Press, (New York), 1990.
29. Cyrac Peter A, Vimalan M, Sagayaraj and Madhavan Physica B: Condensed Matter 405 (2010) 65-71.
30. Datta A and Pati S.K, The Journal of Chemical Physics. 118 (18) (2003) 8420-8427.
31. Dhanaraj G, Byrappa K, Prasad V and Dudley M. (2010). Springer Handbook of Crystal Growth, 3-16.
32. Dhanushkodi J and Ramajothi, Journal of Crystal Research Technology. 39 (2004) 592-597.
33. Domingos S.R, Silva P, Buma S.P, Garcia W.J, Lopes M.H, Paixao N.C and Woutersen S, Journal of Chemical Physics, 136 (13) (2012)134501-134509.
34. Domingos S.R, Silva P.S.P, Buma W.J, Garcia M.H, Lopes N.C, Paixao J.A, Silva M.R and Woutersen S, Journal of Chemical Physics 136 (2012) 134501-134509.
35. Onitsch. E.M, Microscope 95 (1950) 12.
36. Endredi H, Billes F and Holly S, Journal of Molecular Structure. 633 (2003) 73-82.
37. Fleck M. and Petrosyan A.M, Springer, Dordrecht, Journal of salts and amino acids (2014) 1-574.
38. Franken P.A, Hill A.E, Peters C.W and Weinreich G. (1961). Journal of Physical Review Letter 118-119.
39. Frisch M.J, Trucks G.W, Schlegel H.B, Scuseria G.E, Robb M.A, Schlegel H.B, Scuseria G.E, Robb M.A, Cheeseman J.R, Montgomery, Vreven T, Kudin K.N, Burant J.C, Millam

J.M, Iyengar S.S, Tomasi J, Barone V, Mennucci B, Cossi M, Scalmani G,.Rega N, Petersson G.A, Nakatsuji H, Hada M, Ehara M, Toyota K, Fukuda R, Hasegawa J, Ishida M, Nakajima T, Cheeseman J.R, Honda Y, Kitao O, Nakai H, Klene M, Li X, Knox J.E, Hratchian H.P, Cross J.B, Adamo C, Jaramillo J, Gomperts R, Stratmann R.E, Yazyev O, Austin A.J, Cammi R, Pomelli C, Ochterski J.W, Ayala P.Y, Morokuma K, Voth G.A, Salvador P, Dannenberg J.J, Zakrzewski V.G, Dapprich S, Daniels A.D, Strain M.C, Farkas O, Malick D.K, Rabuck A.D, Raghavachari K, Foresman J.B, Ortiz J.V, Cui Q, Baboul A.G, Clifford S, Cioslowski J, Stefanov B.B, Liu G, A.Liashenko, Piskorz P, Komaromi I, Martin R.L, Fox D.J, Keith T.,.Al-Laham M.A, Peng C.Y, Nanayakkara A, Challacombe M, Gill P.M.W, Johnson B, Chen W, Wong M.W, Gonzalez C and Pople J.A, Gaussian 03, Revision C.02, Gaussian Inc.,Wallingford, CT, 2004.

40. Fukui K, Theory of Orientation and Stereo selection, Springer-Verlag, Berlin, (1975).
41. Fukui K, Yonezawa T and Shingu H, Journal of Chemical Physics. 20 (4) (1952) 722-725.
42. Glendening E.D, Landis C.R, Weinhold F, Comput. Mol. Sci. 2 (2011) 01–42.
43. Glendening E.D, Reed A.E, Carpenter J.E, Weinhold F, NBO Version 3.1, Gaussian Ins., Pittsburgh, (2003).
44. Gopalan R.S, Kulkarni G.U and Rao C.N.R, Journal of Chem Phys Chem 1 (2000) 127-135.
45. Guilbert L, Salvestrini J.P, Fontana M.D, and Czapla Z, Journal of Physics Review B, 58 (1998) 2523-2528.
46. Gunasekaran S, Anbalagan G and Pandi S, Journal of Raman Spectroscopy Journal of Raman Spectroscopic. 37 (2006) 892-899.
47. Hemaraju B.C, and Gana Prakash P, Journal of Optik. 129 (2015) 3049-3052.
48. Hubert Joe I, Kostova I, Ravikumar C, Amalanathan M and Pinzaru S.C, Journal of Raman Spectroscopic. 40 (2009) 1033-1039.
49. Hubert Joe I, Kostova I, Ravikumar C, Amalanathan M and Pinzaru S.C, Journal of Raman Spectroscopic. 40 (2009) 1033-1038.
50. Indumathi P, Chitravel R, Saravanan R, Vinitha G and Indra K, Journal of Materials Research innovations (Taylor and Francis). 8 (2017) 396-403.
51. Jyoti Dalal and Binay Kumar, Journal of Optical Materials 51 (2016) 139-147.
52. Karabacak M, Karaca C, Atac A, Eskici M, Karanfil A and Kose E, Spectrochimica Acta Part A: Journal of Molecular and Biomolecular Spectroscopy. 97 (2012) 556-567.

53. Kishan Rao K, and Sirdeshmukh D. B, Bulletin Journal of Materials Science 5 (5) (1983) 449-452.
54. Kleinman D. A. Nonlinear Dielectric Polarization in Optical Media. Physical Review, 126 (6) (1962) 1977-1979.
55. Krishnakumar V and John Xavier R, Journal of Pure Applied Physics 41 (2003) 95-99.
56. Kurtz S.K, Perry T.T, Journal of Applied Physics. 39 (1968) 3798-3813.
57. Lewis D.F.V, Loannides C and Parke D.V., Journal of Xenobiotica 24 (1994) 401-408.
58. Luque F.J, Lopez J. M and Orozco M, Theoretical Chemistry Accounts: Journal of Theoretical Chimica Acta. 103 (3-4) (2000) 343-345.
59. Meyer E, ver Z, Dtsch. Ing. 52 (1908) 645-654.
60. Philip M. Sargent Journal of Materials Science Letters 8 (1989). 1139–1140.
61. Mitchell B. S, John Wiley & Sons. INC. publication, New Jersey (2004).
62. Mohana Priyadarshini K, Chandramohan A, Anandha Babu G and Ramasamy P, Journal of Optik 125 (2014) 1390-1395.
63. Mohd Faizana, Mohammad Jane Alama, Ziya Afrozb, Vítor Hugo Nunes Rodriguesc and Shabbir Ahmada, Journal of Molecular Structure 1155 (2018) 695-710.
64. Muhammad Khalid, Akbar Ali, Muhammad Adeel, Zia Ud Din, Nawaz Tahir, Edson Rodrigues Filho, Javedlqbal, Muhammad Usman Khan, J. Mol.Struc., 1206 (2020) 127755.
65. Mullin J.W and Ang H.M. Nucleation characteristics of aqueous Nickle Ammonium sulphate solution. Journal of Faraday discussions of the Chemical society, (1976) 141-148.
66. Murray J.S and Politzer P, The Journal of Organic Chemistry, 56 (23) (1991) 6715-6717.
67. Nelson D.F, Kleinman D.A, and Wecht K.W, Journal of Applied Physics Letters 30 (2) (1977) 94-96.
68. Nora Okulik and Alicia jubert H, Internet Electronic Journal of Molecular Design 4 (2005) 17-30.
69. Padmaja L, Ravikumar C, Sajan D, Joe I.H, Jayakumar V.S, Pettit G.R and Neilsen F.O, Journal of Raman Spectroscopic 40 (2009) 419-428.
70. Pahurkar V, Anis M, Baig M, Ramteke S, Babu B and Muley G, Journal of Optik. 127 (2016) 4932-4936.

71. Pathak P, Feng H, Hu X, and Mohapatra P, IEEE Communications Surveys & Tutorials 17 (4) (2015) 2047-2077.
72. Patil P.S, Dharmaprakash S.M., Hoong-Kun Fun and Karthikeyan M.S, Journal of crystal Growth 297 (2007) 111-116.
73. Pecaut J and Masse R, Journal of Material Chemistry. 4(12) (1994) 1851-1854.
74. Peter Politzer and Jane Murray S, Journal of Teoretical Chemistry Accounts: 108 (3) (2002) 134-142.
75. Peter Politzer, Donald G, Truhlar, Journal of chemical Applications of Atomic and Molecular Electrostatic Potentials (1994) 1-119.
76. Petrosyan H. A, Karapetyan H. A, Antipin M.Yu and Petrosyan A.M, Journal of Crystal Growth 275 (2005) 1919-1925.
77. Prasad P.N and Wollians D.J, Introduction to Nonlinear Optical Effects in Organic Molecules and Polymers, Wiley, New York, 1991.
78. Preema C. Thomas, Jolly Thomas, Packiam Julius J, Madhavan J, Selvakumar S and Sagayaraj P. Journal of Crystal Growth 277 (2005) 303-307.
79. Premkumar S, Rekha T.N, Mohamed Asath R, Mathavan T and Milton Franklin Benial A, European Journal of Pharmaceutical Science 82 (2016) 115-125.
80. Pucceti G, Perigaud A, Badan J, Ledoux I. and Ziss J, Journal of Optical Society of Americam B 10 (1993) 733-744.
81. R.M. Silverstein, F.X. Webster, John Wiley and Sons, (New York) 2005.
82. Rajagopalan N.R, Krishnamoorthy P, Jayamoorthy K, Journal of Inorganic Organo metallic Polymers and materials 27 (2017) 739-756.
83. Ravikumar C, Joe I.H and Jayakumar V.S, Journal of Chemical Physcis Letter 460 (2008) 552-558.
84. Razzetti M, Ardoino L, Zaotti M, Zha and C. Parorici, Journal of Crystal Research Technology 37 (2002) 456-465.
85. Reed A.E, Curtiss L.E, Weinhold F, Chem. Rev., 88 (1988) 899-926.
86. Roeges N.P.G, A Guide to the Complete Interpretation of Infrared Spectral of Organic Structures,Wiley, Newyark (1994).
87. Sabir H, Mashraqui , Rajesh S. Kenny , Shailesh G. Ghadigaonkar, Anu Krishnan, Mily Bhattacharya , Puspendu K and Das, Journal of Optical Materials. 27 (2004) 257-260.

88. Santhakumari R and Ramamurthi K, Spectrochimica Acta Part A: Journal of Molecular and Biomolecular Spectroscopy 78 (2011) 653-659.

89. Saral H, Ozdamar O, Uçar I, Bekdemir Y, and Aygun M, Journal of Molecular Structure (2016)1103-1109.

90. Saravanan M, Journal of Optical Material 58 (2016) 327-241.

91. Sarojini K, Krishnan H, Kanakam C. and Muthu S, Spectrochimica Acta Part A: Journal of Molecular and Biomolecular Spectroscopy,108 (2013) 159-170.

92. Senthil S, Pari S, Sagayaraj P and Madhavan J, Journal of Physical B. 404 (2009) 1655-1660.

93. Senthil S, Pari S, Xavier R and Madhavan J, Journal of Optik 123 (2) (2012) 104-108.

94. Shaikh M, Anis M, Shirsat M and Hussaini S, Journal of Optik 154 (2018) 435-440.

95. Shanthi A, Krishnan C and Selvarajan P, Journal of Physica Scripta. 88 (2013) 035801-035805.

96. Silverstein R. M, Webster F.X, Spectrometric Identification of Organic Compounds, John Wiley and sons, New York, 2003.

97. Silverstein R.M, F. X. Webster, John Wiley and Sons, New York, 2005.

98. Sjoberg P and Politzer P, Journal of Physical Chemistry. 94 (10) (1990) 3959-3961.

99. Smith B.C, Infrared Spectral Interpretation a Systematic Approach CRC Press Washington, DC, 1999.

100. Socrates G., Infrared and Raman Characteristic Group Frequency, third ed., Wiley, New York, 2001.

101. Spire A, Bathes M, Kellouai G and Nunzio De, Physica D. 137 (2000) 392-401.

102. Stuart B, Simoni-Wastila C, Zuckerman L, Davidoff I. H, Shaffer A, Yang T, Bryant-Comstock L, The American Journal of Geriatric Pharmacotherapy 8 (5) (2010) 441-453.

103. Subhadra K. G, Kishan Rao K & Sirdeshmukh D.B, Journal of Bulletin of Materials Science 23 (2) (2000) 147-150.

104. Sun W, Barant A and Kanalkaran, Journal of Electrochemical Acta 50 (2005) 3359-3374.

105. Sun Y, ChenX, Sun L, Guo X and Lu W, Journal of Chemical Physics Letter. 381 (2003) 397-403.

106. Swaminathan J, Ramalingam M, and Sundaraganesan N, Spectrochimica Acta Part A: Journal of Molecular and Biomolecular Spectroscopy, 71 (5) (2009) 1776-1782.
107. Szafran M, Komasa A, Bartoszak Adamska, J. Mol. Struct., 827 (2007) 101.
108. Tagmatarchis N, Aslanis E, Prassides K, and Shinohara, Journal of Chemical Material 13 (2001) 2374-2379.
109. Tanak H, Journal of Physical Chemistry A. 115 (47) (2011) 13865-13876.
110. Varsanyi G, Baitz E, Billes F, Grofcsi A, Horvath G, Jalsovszky G, Keresztury G, Kiss A, Szoke S, Sztraka L and Toth A, Journal of Acta Physica Academiae Scientiarum Hungaricae, Tomus 35 (1974) 219-238.
111. Varsanyi Assignments for Vibrational Spectra of Seven Hundred Benzene Derivatives, Academia kiado, Budapest, 1973.
112. Vengatesan B, Kanniah N, Ramasvamy P, Journal of Material Science Letter 5 (1986) 987.
113. Vimala M, Journal of Materials Science: Materials in Electronics 28 (2017) 5154 -5164.
114. Weinhold F, Landis C.R, Valency and Bonding: A Natural Bond Orbital Donor–Acceptor Perspective, Cambridge University Press, Cambridge, (2005).
115. Whiffen D.H, Jonural of Spectrochimica Acta 7 (1956) 253-263.
116. Xitao Liu, Xinqiang Wang, Xin Yin, Shande Liu, Wen He, Luyi Zhu Guanghui Zhang and Dong Xu, Jonural of Crystal Engeneering Communication 16 (2014) 930-938.
117. Yugandhar P, Vasavi T, Jayavardhana Rao Y, Uma Maheswari Devi P, Narasimha G and Savithramma N, Journal of Cluster Science 29(4) (2018) 743-755.
118. Zhang H.S, Chen G.H and Wang, Jonural of Research of China 20 (2004) 640-648.

CPSIA information can be obtained
at www.ICGtesting.com
Printed in the USA
BVHW052113110423
662150BV00014B/541